For the Beauty of the Earth:
Solutions to NET ZERO ENERGY

FRANK C. PAO

Archway Publishing books may be ordered through booksellers or by contacting:

Archway Publishing
1663 Liberty Drive
Bloomington, IN 47403
www.archwaypublishing.com
844-669-3957

ISBN: 978-1-6657-3964-1 (sc)
ISBN: 978-1-6657-3962-7 (hc)
ISBN: 978-1-6657-3963-4 (e)

Library of Congress Control Number: 2023903748

Print information available on the last page.

Archway Publishing rev. date: 10/09/2023

I would like to dedicate this book to my family and to all parents and grandparents who worked tirelessly for this cause.

Contents

Preface

The purpose of writing this book is my concern for the younger coming generations. What kind of world are we leaving them. In light of the rate that climate change is progressing, to this day, there is little being done while opposition abounds against the changes that need to be done. In addition, with so many countries still dependent on fossil fuel, the war in Ukraine is causing devastation and human suffering which is totally uncalled for. In recent years many experts on renewable energy and climate change believe that solar energy should provide 40 percent of all the energy. To this day, the world has only got 1 percent installed. So there is long way to go to meet 2050. We had better get it going in the coming years.

The title of the book comes from the hymn "For the Beauty of the Earth," which details the wonderful features and lives on earth. We have inherited a beautiful planet, and we should all try to treasure it as well as preserving it. However, we have spent far more resources on weapons to destroy ourselves and planet Earth, but we are reluctant to work together and save ourselves. Can we afford this anymore?

It is always better to make friends than enemies. In my personal experience coming to America, it was wonderful to have the opportunity to meet students from all around the world; it was very enjoyable, and later on many became my friends. In addition, a number of years ago, there was a very interesting movie, *Joyeux Noel*, where soldiers in their trenches in World War I celebrated Christmas and later on became friends with their enemies. Why can't we do the same? We should work together with all countries in the world as one in the effort to save us all.

In addition, considering the situation of Russia and Ukraine, fossil fuel and nuclear power plants play a major role that is very concerning. However, applying renewables like solar and wind to a distributed generation system like a microgrid would have made a sizable difference. Meanwhile, much human suffering and destruction could have been avoided.

The Paris Accord was a great start but needs mandates on commitments. Mandates are vital if we are going to solve the problems. Time is of the essence. On mandate, I would like to give a special thanks to Governor Charlie Baker of Massachusetts for signing the January 2021 bill on net zero energy by 2050, joined by Senator Marc Pacheco, who was called the North Star of this project; Minority Leader Bradley Jones, and Senator Patrick O'Conner, who were instrumental in making it a reality. Consequently, Massachusetts now leads the nation on clean energy legislation and hopefully more states will come on board.

Dr. Patrick Hofer of 3S Swiss Solar Solutions continues to spearhead the advancements and broaden the application and the quality of BIPV (Building Integrated Photovoltaic) products. Joe Morrissey has advised on implementation in handling BITERS (Building Integrated Thermal Electric Roofing Systems) from the beginning. Recently, I thank the MIT Climate CoLab for the invitation to be a member. This organization connects key players, especially scientists, throughout the world in solving climate change, together with the United Nations. Yes, we all have to work together to handle the problems ahead of us all. Many thanks to President Biden, Vice President Harris, Speaker Pelosi, and Majority Leader Schumer; to Senators Markey, Warren, and Sanders and to Representatives Jayapal, Ocasio-Cortez, and Porter, together with the rest of Congress in successfully passing the Inflation Reduction Act that is a monumental achievement.

To anyone interested in learning more the technical side of solar technologies, especially in the BIPV areas, please look into *Building Integrated Photovoltaic Thermal Systems: Fundamentals, Design and Applications* by Huiming Yin, Mehdi Zadshir, and Frank Pao published in 2021 by Elsevier Academic Press.

Moving ahead on this situation, we should also look at important quotations: "The only thing necessary for the triumph of evil is for good man to do nothing" (Edmund Burke, 1797).

"We choose to go to the moon in this decade, not because they are easy, but because they are hard, because that goal will serve to organize and measure the best of our energies and skills, because that challenge is one that we are willing to accept, one we are unwilling to postpone and one which we intend to win" (John F. Kennedy, 1962).

Please take a look at the President Kennedy's commencement speech at Rice University. It is very inspiring that he managed to unite the country and got us to the moon..

"Climate change is not a project for a few individuals or a few organizations or a few nations. It is a global project. There are great technologies available and sharing them appropriately could be mutually beneficial and expedite the project. Time is of the essence."

As Greta Thunberg mentioned at the UN, large corporations care about profits above all. Profits are important, but our future livelihood is way more vital. There should be a solution to balance both sides and move ahead. It is crucial that more large corporations are willing to make aggressive moves on investing in quality, cutting-edge sustainable technologies. Legislators must make appropriate guidelines to their states and cities as well as assurance that they are being observed; otherwise they will not achieve results.

One of the major renewable energy technologies is solar or photovoltaic. Potentially, it will be the largest sector of renewable energy. Though there are numerous solar technologies, the main emphasis would be on silicon crystalline technology because it is the most abundant.

However, despite its being so plentiful, currently, it is not being applied wisely. Quite a sizable amount of standard solar panels built are not up to par on quality, which sadly lands them in landfills after a short few years after installation. It not only affects the owners; it also hurts the banks as well. Yes, cost is important, but quality is paramount because it is an energy generation product which is being relied on all the time. Building Integrated Photovoltaic (BIPV) products are the way to go. Though this is a solar panel, it is also part of the weathering skin of the building—e.g., roof tile, skylights, atriums, shades and façades. They usually have to be sturdily built because they are part of the building. In addition, they are better protected in adverse weather conditions.

Solar cell development on conversion efficiencies have been making advancement throughout the years. Unfortunately, it has been relatively incremental. Therefore, adding a thermal system beneath a solar roof tile, which would more than double the solar conversion efficiency, is an option for many locations.

Around fifteen years plus ago, our organization developed the hybrid system known as BITERS (Building Integrated Thermal Electric Roofing System) (7). Right beneath the PV roof tile is the thermal system with tubing circulating liquid which extracts the thermal energy from the roof. This energy can be used for domestic applications, storing excess heat as well as reducing the air conditioning cost in the summer. The thermal system cooling also managed to maintain the efficiency of the solar cells because the efficient starts degrading as the temperatures reach 85 degrees F., by up to 10 percent as it reaches 110 degrees F.

Meanwhile, this could also extend the life of the solar cell. Currently, there are numerous installations in the northeast, mid-Atlantic states, and Midwest. All of them are working well.

The BITERS has a conversion efficiency ranging from 38 to 50 percent, subject to location. Considering the solar energy conversion efficiency that has been achieved, the houses that had installed the product managed to achieve net zero energy or be totally self-sustaining. Nowadays, as weather conditions grow more severe, less reliance on the grid is recommended. As to cost, there were questions such as why not put more panels on the roof to get more energy. Unfortunately, most slope roofs do not offer space for the panels to obtain the energy.

Acknowledgement

Thanks to Joe Morrissey of Aesthetic Green Power Corporation for some of the images. Sacramento Municipal Utility District (SMUD) pioneered the solar program. It was in time for Atlantis to deliver the newly developed BIPV solar roof tile Sunslate to the housing project into the district.

Thanks to Prof. Huiming Yin of Columbia University, New York, for advice on technical issues; to Prof. Anton Falkeis and Architect of University of Applied Arts in Vienna, Austria, who provided information on the Active Energy Building; to Mr. William James, CEO of Jpods, who provided information on personal rapid transit; and to Mr. Sylvain Cote, architect and builder who creatively turned an eighty-three-year-old house into "The Beauty and the Beach" by applying the most advanced green technologies.

Thanks to Mr. Jonathan Phillips, Managing Director of Anka Funds, for recommending Atlantis to install the BITERS system on the Cherokee Home as well as providing information on the project, and to Mr. Phillip Schneider of Creative Energy Solar in Wyoming for providing images of the Welcome Center. Also, thanks to Dr. Todd Duellman for providing information on his net zero energy home using the BITERS. Thanks to Como Park office on the images.

Thanks to Mr. George Bialecki, President of Alternative Living Foundation; Dr. Tao, Prof. of Florida International University; and the late Louis Demitri who all worked tirelessly in making Future House USA a great success.

Thanks to Dr. Richard Perez of the Energy Source image and thanks to Lajos Heder on the Austin Sunflower. On the first net zero energy home in Montana, thanks to Lee Tavenner for the information

Thanks to Lee Tavenner of Solar Plexus on the image and the design of the NET ZERO ENERGY Sunslate system.

Finally, I would like to thank the staff of Atlantis, 3S, and Golden Solar, who worked tirelessly to make the solar power system, especially the BIPV modules on the Water and Life Museum in Hemet, California, which earned the Iconic Award from Public Architecture with Distinction by the German Design Council in 2016. Many thanks to Tudor Galetive in managed to get the UL Certification on time. Thanks to Marc Schindler, Managing Director of Ottobahn GmbH. for the image and write-up of their rapid transit system. In addition, thanks to Greg Mills and Carl Miller for the images of TallSlate on their homes.

Reference:

https://www.youtube.com/watch?v=QXqlziZV63k (1962) (Oct. 2022)

Chapter 1

INTRODUCTION

The Reason for Writing This Book

The main reasons for writing this book are our concerns over climate change, energy sustainability, achieving net zero energy on solar technology applications. An additional concern is the planet Earth that we will be leaving for the coming generations. The concern of the younger generations on climate change is key. *Lancet*, December 2021) (2) and the *New York Times* through the Pew Research Center in March 2022 (3) together with the Pew Research Center Report of May 2021 on Generation Z and Millennials, show that they have been active in supporting renewables but stopped short of a break from fossil fuel (4).

Solar is one of the major renewable resources. According to the Pew Research Center report (3), only 30 percent of the population would be interested in total renewables. Fossil fuel has been used for over a century. Changing paradigms is always a big hurdle unless there is strong show-and-tell. Throughout so many decades, reliance on energy from utilities has been the norm; change would not comfortable to many. However, now is the starting point to learn, due to our own energy security. In light of the digital age one should be concerned about attack from our adversaries or horrific storms, when a sizable amount of people will have no power—especially in the winter, such as happened to Texas in February 21.

Dr. Stephen Chu, former Secretary of Energy, has strongly recommended more insulation on our buildings and distributed power generation because transmission lines are costly. Back in 2008, power line installation cost $ 3.0 million/mile. On top of that, substantial energy is lost during transmission. Recently, considering the situation in Russia and Ukraine, fossil fuel and nuclear power plant play a major role that is very concerning. However, in that case renewables like solar and wind, as well as distributed generation like a microgrid, would have made a sizable difference.

Today, most solar panels are made from silicon, which is the second most abundant element in the world. Why don't nations large or small start looking into producing their own silicon? Iceland and Taiwan produce their own silicon. There is no reason why more countries cannot do the same. The result would be more energy independence and more stability in all our societies, where we could be more friendly to one another instead.

On solar, to achieve net zero energy, Building Integrated Photovoltaic (BIPV) as well as adding thermal on the same roof panel is one of the solutions. Product quality is paramount because it is an energy generation system. Meanwhile, despite so many international conference actions, there is no real serious mandate commitment by major nations to take an active role on climate change issues, and time is running out to turn the tide. Considering the scarcity of resources, it is vital we should look into building strong, well-insulated, and energy-efficient buildings that can withstand high winds, snowstorms, and fire. Since the initial installation of Sunslate, which is the one of the first BIPV roofing products, it has encountered hardly any issues with adverse weather conditions for almost twenty years with installations in Montana; Aspen, Colorado; Jones Beach, New York; Orlando, Florida; and Jamaica. They all went through heavy snow storms and hurricanes. Currently, they are all in great condition. Not only it is rugged; it is also pleasing aesthetically. To this day, none of the installations has a single leakage. All these installations could probably last for another twenty to thirty years or more.

Though the BITERS have managed to get such great conversion efficiency, installation of the system in the field is time-consuming, and the likelihood of making errors is quite high. To prevent or minimize the chances of error, the modular BITERS system was developed, which makes installation a lot simpler as well as faster. The modular BITERS system is assembled and tested at the plant before shipping out. Once it arrives in the field, it is hoisted by a crane onto the roof. Each modular system is mounted on a roof and can be easily connected.

Considering the performance of the BITERS (7) system, we have come up with a mobile design which can be used for emergency relief situations, especially in remote areas where energy systems are not available or damaged. It can be used for communication, lighting, water purification, or desalination.

On the glass BIPV glazing side, it can be used in façades, shades, canopies, greenhouses, and skylight projects that Atlantis has installed. Most of them are ten to fifteen years old, and to this day, they have weathered severe storms and are still operation according to specifications. BITERS can also be implemented on façades or shades as well. In addition, there are special projects such as Sunflower co-ops in Austin, Sunsail shades, and solar trees. Most of these projects will be presented in this book. In tackling climate change issues, building better houses and buildings is vital. They should be able to withstand severe storms, flooding, and fire as well as being able to sustain the energy throughout any given period. Insulation is also very critical.

Window technologies have come a long way, and applying them to new buildings would be advantageous. Just look into other countries, especially Europe, and see to how their buildings are more energy-efficient and can withstand storms and fire. BIPV is opening a new area of building materials. It could also be called ABUM (active building material) or ABUP (active building product) (8) with no moving parts. As we get into more applications successfully, the volume of production will increase and become more widely recognized. The price will come down, which will result in attracting more installations—the sooner the better, and the earlier climate change issues will be solved.

Throughout the years living in upstate New York in the seventies, eighties, and early nineties, it was very enjoyable driving along the Hudson River, Taconic Parkway, and part of Route 9. Mother nature is beautiful. Given such a great privilege of inheriting the earth, for so many years we have not kept pace in taking care of it, which is shameful. These days, driving through the area raises climate change concerns and doubts of how long the beautiful scenery will be around. Still, for the past three decades or so, advancement on renewable technologies like solar and wind has been quite substantial, but serious implementations have been slow.

Our planet's current situation is alarming. According to the UN IPCC, if no drastic action is taken in the next ten years, the climate change situation will not be reversible, which would make a great portion of our planet uninhabitable and seriously threaten the food supply. Meanwhile, countless plants, wildlife animals, birds, and insects that are important to our ecosystem will disappear. In addition, food growth could be a major problem due to extreme weather situations. Right now, farming has already occupied more than 40 percent of the land. Greenhouses would help a lot, but using the land wisely is critical.

Coal, which was most commonly used throughout the past century, is still the prime resource in generating our electricity to this day. We all know that electricity is the most critical energy in modernizing an economy.

Unfortunately, burning coal to generate electricity is a polluting process and a very limited resource. In addition, it is hazardous to mine and difficult to handle. Despite all these concerns, coal plants are still operating.

Oil is also being used to generate electricity, along with many other applications, including automobile fuel, medicine, plastic, and fabrics. North America at one point was the largest producer. Most of the wells are dried up already, and North America is now growing more dependent on imports, especially from Mideastern countries, which are quite hostile. It is a bad polluter, demand continues to rise due to the rapid growth of economies like China and India, so supply is going to be a major problem. Today oil rigs are going down more than five thousand feet beneath the ocean, which clearly shows that the days of cheap oil are over. Currently, a lot of oil in the US is produced through fracking, which is environmentally hazardous.

Natural gas is probably the cleanest of all extracted products listed so far but is still an emitter of carbon. However, transportation can also be a problem. Though there is still a sizable supply available, it may not last for a reasonable long time. The current fracking process, which is how companies are extracting it, is a concern to the environment. It is affecting and contaminating the underground water, which could be a health hazard.

Nuclear power has seen renewed interest because it does not produce carbon by-products. However, it could pose grave concern due to national security. It is hazardous, regardless of advocates claiming that it is safe. Despite safe operation of the system, dealing with waste in future years will be an enormous task. It is time-consuming and expensive to build a nuclear power plant, and the amount of resource is still limited. Back in the summer of 1973, staying at Fletcher Hall, I listened to some of the students' concerns about nuclear reactors. They strongly suggested that solar energy could be the solution. At that time the price was way too high and it was on the back burner at that time.

In addition, water is already a scarce resource, and it is vital to our very livelihood. However, though a sizable portion of water is being consumed in generating hydroelectric energy, we should find ways to reuse the precious commodity. It is imperative to conserve this precious resource for many other applications.

Looking at all the resources above, why not consider the ones that are clean and abundant, which could last us for centuries to come? Next to oxygen, silicon is the most abundant resource in the world. Yet no major effort has gone to implementing silicon solar technology in a massive way. It is clean, without moving parts, durable, reliable, and not toxic.

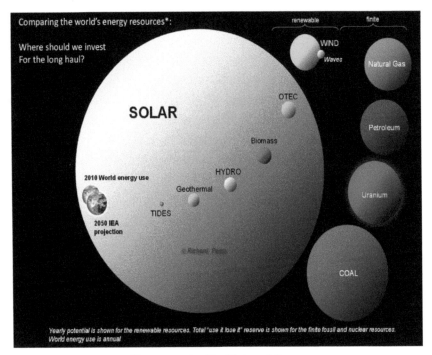

Courtesy to Dr. Richard Perez

Aside from energy, other resources must be considered such as building materials, metal, chemicals, wood, and all nonrecyclable materials. As our population grows, all these resources will be less available if they are not conserved wisely. Batteries are also a concern, especially lithium-ion batteries in automobiles, which can deliver from 500 to 1200 cycles. Currently, the supply of lithium worldwide is limited to a small number of countries. Disposing of them could be a problem as well. Meanwhile, we could look into development of batteries to give 50,000 to 100,000 cycles.

One promising prospect is transitioning to fuel cells, which could generate energy through combustion of hydrogen and oxygen. which are abundant, and the process generates no carbon pollution. Fuel cells are quiet and reusable. Though the technology is not quite here yet due to cost, it is getting close. A lot of organizations are working diligently in this area. The current fuel cell applications on automobiles in the market could go around 300,000 to 450,000 miles, which is coming close to internal combustion engines. As further advancements go, there is a great possibility that it could go beyond 600,000 to a million miles or more, a life span improvement as compared to the internal combustion engine.

Back in the eighties, in the course of planning to build a house in a new development in Upstate New York, the developer would not allow solar panels on the roof. Reluctantly, an agreement was made. In the late eighties, there were thoughts of a solar roof tile to be built. In the early nineties, a search was made around the Hudson Valley to see whether a tile was available that could be a suitable solar roof tile. Unfortunately, it was not available. In 1998, at a solar conference and exhibition in Vienna, it was possible to locate a couple of companies that had started doing solar roof tiles. One was Canon of Japan and the other was Atlantis of Switzerland. After going through the products, we decided to learn more on the Atlantis product. In 2000, after visiting the operation, a decision was made on the company. After the annual general meeting, the board members and shareholders decided to reorganize the company. The Swiss operation was renamed 3S (Swiss Solar Systems AG) with CEO Dr. Patrick Hofer. The US operation was named AES (Atlantis Energy Systems Inc.). Atlantis took over the Sunslate solar roof tile product.

Initially, there were some problems, but they were later solved. Since then, all the installations to this day have been successful, with some minor hiccups. There has been not a single leakage despite harsh weather conditions throughout all these years. While the Sunslate is the main product of the company, Atlantis was asked to consider manufacturing custom modules for facades, skylight, greenhouses, and other applications. Then we began realizing that we were actually opening a completely new field and trademarked Active Building Material (ABUM) and Active Building Products (ABUP).

While developing the BITERS, Atlantis was invited to participate in deploying in a number of locations almost simultaneously: the Future House USA in Beijing, China; the Cherokee Home in Raleigh, North Carolina; and the Sullivan Estate in Hawaii.

The BITERS has been here for more than ten years, and all the installations are still in great conduction. All of them managed to survive harsh or extreme weather conditions. So far, a large portion of the installations on rooftops are standard modules, not a sustainable arrangement. Though they are relatively easy to install, they could create lot of problems in extreme weather conditions, especially leakages, because they have penetrations on the roofs. Aesthetically, they not pleasing, and in addition, most of them are not well made due to materials sourced from cheap imports. The main reason that Building Integrated Solar Technology has not been popular is due to builders and contractors not willing to install them unless the customers insist.

Houses: Major points and components

Location of a house is a critical factor in where one will be spending a great period of one's life. It also has to be suitable to the owner's lifestyle and conveniences as well as the value. As to the plot of land to be built on, is the ground solid, on bedrock, and on a high point? Is it south facing? Waterfront may also be a problem these days. All these are critical points to be considered.

1. Land: Make sure the ground is solid or on the rock because the house is going to be heavy. Especially, if it is on the rock, piling is not needed. Just build a good foundation
2. High point: It is very critical especially on the waterfront. If it is not on the waterfront, look at the potential of flooding.
3. South facing: It is always wonderful to have a south facing house, which can be warmer in the winter and great to apply solar on the roof.
4. Woods and forest: Try to avoid, especially in dry climate areas
5. Distant from dams: Try to be some distance from any dam.
6. Public transportation: If possible, it would be great if is not too far from public transportation.
7. Waterfront locations, especially in front of the ocean, could be very attractive and tempting. However, if it is on higher ground, one could seriously consider it.

These are some of the points that one should consider in locating the land. Often it is not easy. Sometimes the plot of land looks wonderful, but on close look, and considering the area, it could be quite discouraging. One always prefers a flat piece of land, which is easy to build. What happens if it is surrounded by some hills? In case there are heavy storms or rain, flooding could be a problem. it is always good be higher up so that it could have a great view. If the roads are steep and narrow especially in the winter, they can cause serious problem as well as being dangerous.

Construction of the house should be taken more seriously than it has been in the past. It should be well built. All the walls, both internal and external, should be well insulated and fireproof. Well-insulated walls could substantially lower one's utility bill and make it a lot easier to achieve net zero energy or energy independence. To prepare for the extreme weather conditions threatened by climate change, the structure should be able to withstand at least 140 mph wind load or higher and a heavy snow load as well. In anticipation of climate change, it is time to prepare for it with wall thickness.

Houses especially are supposedly our shelter. There is no reason why it is necessary to evacuate when a storm is coming unless there is a serious potential of flooding. All lot of places in the world do not have the luxury of evacuation. The only time to evacuate is in the event of a wildfire. If the building is really well built, most of your belongings will still be there after the fire. We should insist that our government officials mandate such requirements on the building code.

Sustainability is our main objective, and discarding unnecessary material is a big concern, because resources should not be wasted. Finding places to dump them could also be a problem. In the long run, getting a good product is actually less costly. This book gives the general public the basic knowledge of building integrated solar technologies so that all of us appreciate the aesthetic aspects of the installations, their durability, and their energy independence. Quality and sustainability go hand and hand. Good quality products go a long way.

A friend went to study in Paris. She noticed some of her French classmates were carrying some very fancy French handbags and thought they must be really well-to-do. Later on, she found out they had inherited them from their grandmother. Just imagine, when inherited from grandparents, it could remind us some of the fond memories one spent with them while they were around. So it is wise and one of the main reasons not to throw things away. Yes, in changing paradigms on energy from fossil fuel to solar and from centralized to distributed, it is imperative to have quality products that will last for a long time.

House Construction

Quality house construction is vital to achieving net zero energy. In the past, we always looked at the economical ways of managing the house. Cost is always essential, but quality is vital. It could not only provide conveniences but also security and long-term sustainability to us all. For example, a shingle roof, after ten to fifteen years or maybe more, must be replaced. One could use shingles, but a slate roof could last us a lifetime. In addition, we don't have to get contractors to do the work or discard the used shingles into a landfill.

Other examples are walls of brick and stone. They do not need to be painted every now and then. The initial price is a bit on the high side, but these walls can last much long with hardly any maintenance. On top of that, they are better insulators, which could save energy cost. According to the Department of Energy, it is vital to build a house or building that can last for a long time, probably in the order of centuries, like a lot of houses and building in Europe, especially with good insulation. Construction of walls around ten to twelve inches thick could not only last a long time with great insulation; they could also withstand high winds and fire. The building itself could reach the Energy Star HERS (Home Energy Rating System) level at 40 or below, which is very close to net zero energy status.

One of the concerns on construction materials is concrete, which affects greenhouse gas emissions. Concrete consists of cement, sand, gravel, and water. Basically, cement production is where gas emission only consists of less than 20 percent of the concrete mix. However, lately, by using natural gas instead of coal, there have been ways to curb the amount of CO_2 emission by 55 percent, which is quite substantial. Meanwhile, in recent years Marin County in California has set up a low-carbon code that will reduce the carbon emission situation. In addition, there is the surprising discovery that concrete will absorb CO_2 from the air, up to 43 percent of the original emission, in the long term. This may get scientists to believe that concrete is green. Above all, concrete is also a great insulator, which also reduce the energy usage of a building. Consequently, we should all reconsider the application of concrete in building construction (5,6).

Concrete buildings are usually a lot stronger than wood-framed houses. They can withstand fire and heavy wind load situations. All this would be subject to local building code. It is always wise to give a hard look on all these options before buying a home or building a new one. It is a place that will protect us from all kinds of potential adverse situations. Evacuation is not always a good solution. There are lot of places around the world where buildings are so well built that despite heavy storms, the residents are well protected. We should also look into our legislators to make sure that the building codes are up to these standards and that builders adhere to the code. Nowadays, more advanced building materials are available, such as foam concrete or Fox Blocks. ICF (Insulated Concrete Form), with high R factors, can withstand high winds and fire. All these should be considered prior to making any decisions. In light of all these items available to us, our building or house could reach Energy Star HERS 40 or below. As to appliance products, such as heat pumps, which could take care of our house ventilation in hot and cold weather conditions, involve looking into the quality track records together with energy consumptions.

Radiant heat is highly recommended, as well as tilt-and-turn windows which would allow fresh air to come through when the indoor air is stuffy. All these items will help on the indoor air quality, which is vital to our health, work performance, and the well-being of our family. A tankless water heater is also a plus because one does not have to constantly heat the water tank, and hot or warm water can be accessed instantly, which will save water as well. Once having all these in place, one could start to look at energy requirements. It is highly recommended that green hydrogen should be seriously considered as our fuel instead of batteries because it is totally green and renewable.

Wall thickness recommended at the Umwelt Arena

Batteries offer short-term relief. In the long term, hydrogen is the solution. It is the third most abundant element, and it is not toxic. In retrospect, batteries, especially lithium-ion ones, involve several concerns. Organic liquid electrolytes could be volatile and potentially flammable at high temperature. In addition, when the batteries corrode and the chemicals get into the groundwater through the soil, they could contaminate our ecosystems, which would affect our plants and water systems. Landfill fires might result, sometimes smoldering for years to come. All these are harmful to plants, animals, and humans. On top of all, only a small percentage of lithium-ion batteries can be recycled.

On the energy generation systems, quality is paramount. In the past, relying on our utility companies was the norm, and they have been responsible for delivering energy for many years. However, situations have changed. Our utilities are still highly dependent on fossil fuel and growing more vulnerable to adverse weather conditions, hacking, and frequent power outages. It is about time to be more self-reliant and have them as our backup, which will ensure uninterrupted energy conditions as well as contributing to the use of green renewable energy, which will save our environment.

In view of all that, installation of building integrated solar technology is highly recommended. Basically, it forms part of the building envelope and usually securely situated, whether it is a curtain, walls, roof tiles, skylights, atriums, shades, or sound barriers. The system could comprise custom or standard products which could be subject to the manufacturer. On custom design products, there are quite a bit more details that one has to consider because of architect's practices and local building codes. But on the standard BIPV products, most of specifications have been accepted and tested by organizations like UL and IEC to meet the local code requirements. The Sunslate, the TallSlate, and TallSlate Grandee have been successful standard BIPV products of solar roof tiles for a great number of years. The size and the design would be subject to the owner's need and the architect's recommendations. It is just a one-time initial process and it is worth every bit of cost because one has a better idea of what one is getting and the quality of the material, even if it is a bit hard on our pocketbook.

Meanwhile, looking ahead, one would probably avoid a lot of headaches for years to come especially when one installs a BIPVT (Building Integrated Photovoltaic Thermal) system or the BITERS, which could be an energy independent system; in essence the building itself is a microgrid and is grid-tied as well in case of unforeseen situations, provided the system is properly installed.

It was back in 2000 that the Sunslate was installed in this Bolton home. To this day after twenty-some years, there has been no complaint by the owners. Despite all the adverse New England weather throughout all these years, the product has held on quite well.

At that time, the BITERS was not yet available. If it had been available, the whole house could be net zero energy. It was a 9.0 kw Sunslate system. In view of the size of the building, with additional insulation, it is highly possible to achieve a self-sustaining house.

10 kW System, San Jose, CA

On the above 10 Kw Sunslate system, if thermal system is installed, the house could easily be net zero energy because the solar conversion efficiency would be around 50–60 percent, especially in a place like San Jose. It could also include Jacuzzis as well as extended months of using the swimming pool.

BITERS Application on Hydrogen Systems

The above two installations were around in the early 2000s. Back then, the owners converted the collected energy in DC and using inverters to convert them to AC and feed the electricity to the local grid. However, nowadays, considering all the advancements made through the past twenty years on clean energy technologies, one can easily make the home totally self-sustaining as well as using clean energy generation and storage. Hydrogen could be a long-term fuel because it is clean and is the third most abundant element in the world; why is it not being used instead? Using solar energy, hydrogen could be generated through an electrolyzer by splitting water. The hydrogen could be stored in a tank, and energy could be generated through a fuel cell. The prices of electrolyzers and fuel cells have been coming down and getting more affordable. In addition, there are other means of generating clean hydrogen being developed and commercialized. This will be discussed in a later chapter.

Applying BIPVT (Building Integrated Photovoltaic Thermal) or BITERS technology in obtaining green hydrogen to power a home or a building could make it net zero energy. Nowadays, in light of the advancement of electronic control systems, one could easily know how the system is operating, even remotely.

Sunslate award winning installation in Pays de Loure, France, image not shown

The Sunslate solar panels won the "Stars de Metiers 2013 department" prize awarded by the Banque Populaire Atlantique in partnership with Chamber of Trade and Crafts in the category "Global Innovation Strategy" to Thierry Auguste who, with his team, has provided energy solutions involving renewables and solar projects.

TallSlates

These are designed for ultra-high winds and are Dade County certified as well as designed for installation with natural slates. They can either be grid-connected, or their production can be stored on site with the battery solution of your choice.

Courtesy to Greg Mills

Successful installation in Nova Scotia coast of Canada. It is relatively simple to install and in addition, one could pick matching slates to make it more aesthetically pleasing. Though having the images from the customer, haven't got the approval from them that resulted not showing it. The customer has been quite happy with the project.

Courtesy to Carl Miller

To accommodate more options in roofs for homes and other buildings, in working with 3S Swiss Solar Solutions in Switzerland, they have recently given us the color scheme which is shown below.

The color scheme image bottom (Courtesy of Swiss Solar Solutions)

Having color on the solar roof tile or façade will broaden a lot more applications on the building envelope.

The carpark on this office building has solar installation of around 200 kw. Considering the size of the system, with today's storage and generation technologies, together with thermal, it could be a net zero energy office building.

As mentioned, in applying BIPV, one can think of all imaginable kinds of applications. There have been a lot of utility scale installations of solar panels in fields, which are mainly used by utility companies. However, in recent years, in many states, they have stopped allowing large installation of solar panels in fields because they occupy a lot of land, which is quite precious. One potential way to resolve the situation is to consider applications on sound barriers, covers on highways (6) where some European countries are seriously considering it, and large covered carparks.

OWNER: Municipality
SYSTEM: PV Integrated Sound Barrier
POWER: 9.6 kW
LOCATION: Wallisen, Switzerland

The United States has probably far less railways tracks than most European countries, but solar building integrated systems could be installed on the sound barriers on the railway tracks; apart from energy, it could also prevent people or animals from walking too close along the tracks.

The US has plenty of highways, both large and relatively small, that could be considered for utility scale applications, which could double use of the land space. The structures must be sound and able to withstand high winds, heavy snow load, and strong storms However, the covering could improve the longevity of the road surface as well as reduce efforts to plow snow in the winter months, which amounts to value added to the motorway. Germany, Austria, and Switzerland are developing this application (7) As to the sound barriers, they could also be applicable to highways, especially if they go through urban and suburban areas.

Glass module images not should but The Science Center in Vancouver provides younger kids a start in learning about science and technologies so that they can begin to understand and appreciate it at an early age.

Though the solar modules do not have much power, they provide some energy, as does the daylighting.

Aside from clean energy generation, to maintain a healthy environment, one has to consider another important commodity, which is water. To have clean water is a serious concern. In the past, it was seldom an issue. Throughout recent years, pollution caused by fossil fuel through the air, rain, and foreign materials such as plastic have been getting into our water systems, making it costly to have clean water. In addition, deposited waste or landfill affects food farming. In recent years, plastics are appearing in some of our seafood as well.

References:

1. https://www.thelancet.com/action/showPdf?pii=S2542-5196%2821%2900278-3
2. https://www.pewresearch.org/fact-tank/2019/01/17/where-millennials-end-and-generation-z-begins/?campaign_id=54&emc=edit_clim_20220322&instance_id=56394&nl=climate-forward&regi_id=88423053&segment_id=86208&te=1&user_id=9c342a3c5ab9bf73701e10625708f69b
3. https://www.pewresearch.org/science/2021/05/26/gen-z-millennials-stand-out-for-climate-change-activism-social-media-engagement-with-issue/
4. Amer Hamad Issa Abukhalaf M.E. Maryem Kouhirostami M.E., Toward Greener Concrete for Better Sustainable Environment. Academic Letters, Sept. 2021.
5. Harvey C. (2018) Cement Producers are Developing a Plan CO2 Emission. E & E News
6. https://www.pv-magazine.com/2020/09/01/photovoltaics-for-highways/
7. BITERS, US patent no. Pao & Gottlieb 8,196,369 B2, Pao 8,201,382 B1, Pao 8,365,500 B2 Pao 10,958,212 B2, Pao 10,958,212 B2 plus numerous patents pending status.
8. Registered Trade Mark

Chapter 2

MAJOR PROJECTS AND ICONIC PROJECTS

California District Seven Caltrans Headquarters in Los Angeles

Caltrans headquarters: 100 kw BIPV shade structure system which consists of over one thousand 95watt glass/glass modules. The shade modules not only provide electrical energy, they also provide shade, which cools the building and results in reducing air conditioning cost. It was the largest BIPV system in 2004. In 2005 the whole project won the Pritzker Prize, and the system is still in operation today.

San Diego County Operation Center

The COC dining area with 40 large four-point support thick glass/glass modules at 8.5 ft long and 5.5 ft wide, with around 346 watts each in power

The County Operation Center demonstrates San Diego County's commitment to sustainable design. The office buildings were awarded the LEED Gold certification while the Conference and Dining center earning LEED Platinum certification. The Center earned numerous awards for design and sustainability. It was named "Outstanding Governmental Building of the Year" by the California Center for Sustainable Energy and received the AIA California Council's Merit Award for Sustainable Design.

Bern High Rise 200 Kw of Façade PV Facade

A 200 kw BIPV system installed as a façade on a high-rise apartment building located outside of Bern, Switzerland in autumn 2000. To this day, the system is still operating well. If the system had a thermal system beneath the PV system, the potential of being a net zero energy building would be quite high because buildings in some European countries constructed after the late 1980s are highly insulated, as their building codes require. Having a BIPVT could make a sizable difference.

Megaslate in brownish red on top of the emergency building in
Zurich, Switzerland (courtesy of 3S Swiss Solar Solutions)

Looking at the brownish red rooftop building, the color scheme as shown in the introductory chapter matches relatively well as compared with the original rooftop shown from other parts of the building. In BIPV, matching color and product quality is important, together with good system design, which is paramount, especially for energy generation. If a thermal system is installed beneath the Megaslate, a lot more solar energy would be available, with a possibility of net zero energy. In light of Zurich's climate condition, an addition of a geothermal system would be required.

Wyoming Visitors Welcoming Center – Cheyenne

The Wyoming Visitor Welcoming Center installed the translucent glass/glass modules
(courtesy of Creative Energies)

The center introduces visitors to the history of Wyoming with displays on the early-day settlers' life down to modern times. It includes the use of regular solar modules and BIPV glass/glass modules from Atlantis Energy System, Inc.

Awards:

1) Honor Award, Design Excellence, American Institute of Architects, Colorado Chapter, 2015
2) Honor Award, American Institute of Architects, Denver Chapter, 2013
3) Award of Citation, American Institute of Architects, Western Mountain Region, 2013

Sun Sail
Münsingen, Switzerland

This project was installed in June 1999 and inspiration to us that BIPV solar could be applied to all possibilities one could only imagine. Though one can seek out a lot of options, implementation is paramount. In this case, the Sun Sail has been around for more than twenty years and it still in great condition after being hit by numerous major storms throughout the years. It has hundreds of solar cells on translucent modules collecting sunlight and generating energy. This really a combination of art and modern technology at its best.

Sun Sail Construction Münsingen, Switzerland

The design of the project was well thought out. To achieve a sturdy support of the mast structure, three converging steel pipes of 220 mm in diameter and 22.00 meters high are filled in and held together at a space of 1.00 meter by a steel plate.

Sail Design and Structure

Holding the translucent modules is a curved tube of 135 mm attached to the mast together with a horizontal tube, which results in giving the sail a curve and the horizontal lines. The horizontal and vertical metal ropes hold the weight of the modules. Increasing the slight incline of the mast could increase the efficiency performance of the solar modules. As one can see, in addition to such a sturdy construction of the mast that holds the sail, there are three metal ropes holding the sail all three sides.

Solar Modules

The solar modules on the Sun Sail were developed and manufactured by Atlantis Solar Systems AG. They were specially design for this application. Additional material in strengthening the modules were included to the module in lamination as well as using tempered glass. Monocrystalline solar cells were used to assure higher efficiency and the modules were connected in series.

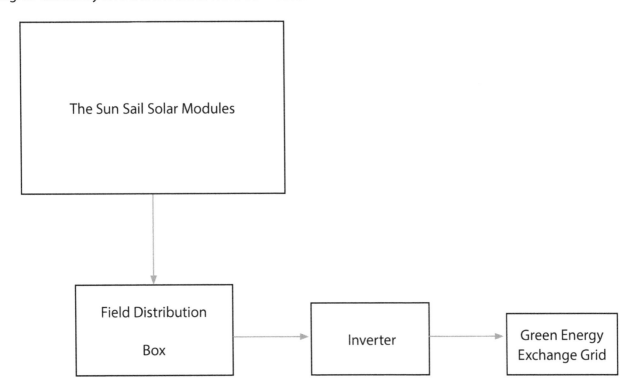

The Sun sail is an 8.2 kw PV system. The project won the Design Competition for the Built Environment Award at the 16th European Solar Energy Conference & Exhibition held in Glasgow, Scotland, in May 2000. Learning from local residences recently, the Münsingen Hospital is planning to replace the solar modules on the Sun Sail because solar cells nowadays have conversion efficiencies of twenty percent plus—that is, almost doubled as compared to those years when the project was installed.

While visiting Switzerland, a short train trip going between Bern to Thun or vice versa, the Sun Sail can be seen from the train; and besides, it is very interesting to spend some time with it.

Shortly after the Sunflower project was installed, we were given this image and it was in the centerfold of the Photon Magazine that drew quite a bit of attention. It is an ironic project located at Mueller Development shopping mail in Austin, Texas along the interstate I-35.

The project consists of fifteen sunflowers which generate a total power of slightly over 15 kw of energy that feed in to the Austin energy grid. In daytime, apart from energy generation, the sunflowers provide some shade to the pedestrians and the bikers. At night, it lights up and provides some lighting on the pathway. In addition, it draws the attention of motorists. It is left to our imagination to envision how one could apply the photovoltaic technologies into these types of creative projects.

Sunflower at night in Austin, Texas (Courtesy of Lajos Heder)

MetLife Stadium Solar Ring

The solar ring at MetLife is an iconic project as well as one of the largest if not the largest BIPV installation in the country. It consists of 1,350 modules with 47 frames, generating 350 kw of clean energy to support 916 LED fixtures around the ring and other areas in the stadium. The lighting reflects from the solar panels at night. As we all know, the MetLife Stadium is the home for the Giants and the Jets. When the Giants play, the LED lights in blue come on, and when the Jets play, the lights are green. The LED lights are also quite versatile. They can be synchronized with music in the stadium, and it can be quite colorful, especially during halftime performances. This illustrates some of the features BIPV technology could make possible. That is probably one of the reasons why MetLife was awarded "Best Stadium Venue of the Year" worldwide at the Stadium Business Summit 2017.

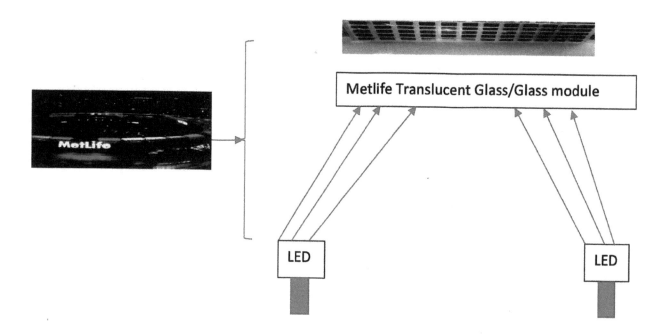

All the modules are glass/glass with relatively thick tempered glass, so that it can withstand heavy wind and snow load or even hurricanes.

One of most important protocols is to make sure that the two modules do not touch each other during installation. All the wiring connections should be properly checked while they are being mounted onto the frames as well.

Climate change is a very serious issue which we have to consider. Though the solar ring system managed to survive the tropical storm Sandy, we should anticipate more severe storms in the coming years, and it is vital that the system will withstand it.

Active Energy Building Vaduz, Liechtenstein

Courtesy to Anton Falkies

This iconic project was designed and developed by the Falkeis Architects in Vienna, Austria. They really look ahead into the future and are able to come up with groundbreaking solutions. The prime objective of the Active Energy Building is to generate more energy than it uses.

There are times when some part of the grid has a problem or neighboring buildings require some energy in emergency situations where it could be helpful. This is one reason for the building-integrated energy production system. The energy supply for this building in Vaduz comes from renewable sources like geothermal energy and from active and passive usage of solar energy. It is an energy-autonomous building using renewable energy exclusively, which results in reducing CO_2 emissions. On the active energy side, the building contains a movable building envelope to obtain solar radiation for energy production applied to heating and radiation for cooling. The building is like a local power plant, utilizing hydro power to temporary store the surplus solar energy generated at peak hours through e-mobility. The whole roof together with the south-facing part of the building uses solar energy with building-integrated PV tracking, which can increase the energy by 2.9 times.

There are seven moving "climate wings" located on the eastern and western parts of the building, applying Phase Change Material (PCM) in latent heat storage. A great portion of the energy is stored in the PCM in the climate wings while exposed to the sun. Later, when the wings are folded, energy will be released for the ventilation in the building. In addition, there are thirteen modules controlled astronomically from east to west. Each of them operates individually in order to minimize mutual shading. In cooperation with Lucerne University of Applied Science, this type of decentralized energy production system has been evaluated by having a full-scale mock-up and is movable to different location where the active energy building can be part of a network, and energy production can be monitored,

The building consists of twelve apartments of 245 square meters each at twenty-one euros per square meter. The objective is to achieve design to allow optimal use of passive solar energy.

Courtesy to Anton Falkies

The climate control system is largely based on geothermal, with the support of seven extendable elements. Three on the east of the façade supply 15 percent of the energy required for cooling, while four on the west side contribute 10 percent of the energy required for heat generation. Phase Change Material (PCM) paraffin is used for latent heat accumulators. It emits a large amount of heat when melting from a solid state to a liquid.

PCM melts at 31 degrees C and solidifies at 21 degrees. Once it is folded, the air circulates through the frozen material, thereby cooling it. The 32 kw PV system in total is designed to generate more power than the occupants consume. The peak yield is not stored in the basement but excess power is shared with neighbors.

The modules on the array installed on the façade changes inclination according to the sun's position and is simultaneously used to generate power and shading. The dual axis tracking is something of a rarity.

All the solar panels can be raised up and outward.

To optimize the passive gains, the architecture team came up with the design tool "solar erosion." Based on the local building code, their input maximized the building volume as well as transferring the numbers topographically so that the structure could get maximum solar exposure in order to get passive gain of the sun's energy.

Awards:

1) Energy Global Award
2) Austria Green Planet Building Award

Future House USA

This is probably one of the first houses using solar technology in the world that managed to achieve net zero energy at the time. Ten countries including China participated at the Olympic Village to demonstrate their latest technologies on alternative energies. In light of the situation, the America House was able to claim the performance of a net zero energy. The solar roofing system installed was state-of-the-art not only because it was able to optimize the usage of the roofing space, but also because it had the thermal system beneath the solar electric solar roof tiles where the combination of both systems in the same roof plain managed to have total energy conversion efficiencies on the order of 30 percent or higher in a place like Beijing instead of only 12–14 percent. In addition, with the thermal system beneath, it cooled the roof, which resulted in maintaining the solar electric and also reduced the air conditioning energy requirement.

Future House USA

1) The solar electric system comes from the Sunslate and the Solar Tree. Sunslate is the solar roof tile installed on the south-facing roof of the America House and is an integral part of the roofing system while the Solar Trees are laid out in the backyard. Both systems generate DC electric power and charged up the batteries. The DC power supports some appliances and lighting. In addition, the batteries are connected to the inverters which support the AC appliances and feed to the grid as well.

2) The solar thermal system is directly beneath the Sunslate system. It consists of thermal tubing with glycol being pumped and circulated throughout the roof, extracting the thermal energy to heat the domestic hot water through the heat exchanger. Considering the size of the system installed, there could be a sufficient amount that could power a swimming pool and Jacuzzi in addition to regular daily requirements.

3) The geothermal system has a heat pump with a geothermal loop buried beneath the ground. It handles the HVAC (Heat Ventilation and Air Conditioning) requirements of the house. In the summer months, the system extracts the heat from the house and stores it in the ground, which results in

cooling the building. In the winter months, it pulls the heat stored from the loop in the ground by circling glycol and warms the house. Consequently, this reduces the size of the system to meet the heating and cooling needs.

All three systems basically compensate each other. With thermal tubing installed beneath the Sunslate photovoltaic system, this allows the system to maintain efficient performance on solar energy conversion in the summer months. In addition, the heat extracted through the thermal tubing from the attic together with the geothermal system could cool the house substantially, resulting in reduced air conditioning requirements.

On top all these systems, the house is constructed with SIP (Structured Insulated Panel) that has very high R factors, which is also vital to energy conservation.

- High Energy Performance - A completely balanced space conditioning system within a high-performance structure, combined to maximize comfort, minimize monthly energy bills, and accentuate environmental stewardship.
- Nontoxic - Only user-friendly, nontoxic materials, adhesives, and laminates are used in FHUSA—no formaldehyde or volatile organic compounds (VOCs). Additional healthy features include kitchen cabinets made from wheat straw, foot pedal controls for faucets in the kitchen and bathrooms, and completely filtered indoor air. The nonpolluting roofing system, combined with the filtering action of the porous pavers, filters all storm water runoff before it returns to the watershed.
- Zero Carbon Footprint - Because the FHUSA creates, on site, all the energy it uses annually, the operational carbon footprint is zero.
- Zero Energy Home – Using the energy stored in the ground under the backyard (ground source energy) and the sunlight falling on the southern roof area (photovoltaic and solar thermal) FHUSA represents the state of the art as a self - sustainable, healthy home system. Officially known as a Zero Net Energy Home.
- The house is designed with connections for electric or hybrid cars in the three cars garage, thus creating a complete 'sustainable home system'.

Future House USA employs an advanced design which integrates the architectural style and innovation of Frank Lloyd Wright with the Chinese practice of feng shui. This approach integrates the home into the rhythms of the surrounding landscape, thus creating a refreshingly smooth flow of air and light.

The Future House USA design represents the latest in sustainable energy and nontoxic technologies; clothed in a tried and true award-winning design style.

Interior space fluidity is enhanced by minimization of interior walls. Open and multitiered interior rooms feature much natural light and regally high ceilings and are all finished with Chinese indigenous materials. Future House USA integrates energy efficiency, indoor air quality, water conservation, municipal water consumption, and indoor air quality into quite literally the house of the future.

George explaining the house system to the US Secretary

After the secretaries got to know the operation of the American House, they complimented that certain parts of the ceilings were a bit high, and they really liked the LED lighting. Actually, Secretary Chu asked about the company or the manufacturer. As a whole, the visit was a great success. However, the Future House USA organization wanted to offer access the performance of the system operation, but it was not able to do so. Though only lead acid batteries were used, it was the best storage that was available, and the control system was still a bit primitive. However, the system managed to achieve net zero energy.

Water + Life Museum Hemet, California

This is one of the iconic projects in which Atlantis participated. The actual project started after the completion of Diamond Valley Lake, which is supposed to be largest man-made lake in the country and could support Southern California in case there is a drought for six month or so. In addition, they built three dams to hold the water. They are East Dam, West Dam and Saddle Dam. It is located between Los Angeles and San Diego. The water comes from the Rocky Mountains through the Colorado River Aqueduct. The solar energy installed in the museums is 550 kw, which covers around 50 percent of the energy requirement. Around 50 kw is the BIPV module system at the entrance of the museums.

During excavation, a sizable number of Pleistocene epoch fossils were discovered, which resulted in a second museum. The two museums are the Center for Water Education and the Western Center for Archaeology. The two museums include laboratories, classrooms, and two auditoriums, together with a number of meeting rooms.

Water & Life Museum Entrance, Hemet, California

As we all know, water is one of the most precious commodities. Though it is abundant, yet a large part of the world has serious water issues. Conservation is paramount, and so it has to be applied wisely to optimize the usage. The Water Museum employs the latest state-of-the-art irrigation drip system. It is also about water conservation. Looking at the twenty-first century, food, energy, and water (FEW) are critical items, a reality which is already hitting us. On food, we should look into multilayer greenhouse because land is a critical item as well. Right now, 40 percent of the land in the world is occupied by agriculture, which is quite substantial. What can we do about it in years ahead? As the population grows, we have to plan the land use wisely in every application with sound environmental planning.

The BIPV Glass/Glass module is a four-point support module and a relatively large one to build. Considering the size and concerns about durability, the technical people developed a new material process, which was a bit challenging to laminate. Thanks to the tireless work of the Atlantis and 3S staff in overcoming it, the design was successful in to passing the Underwriting Laboratories testing protocol to meet the delivery schedule.

Whenever one gets to visit Southern California, it is highly recommended to visit the whole campus of the Water + Life Museum. There is so much to learn about the building design as well as the information provided in the two museums.

Awards:

1) Merit Award for Best/Public Special Use Facility – Pacific Coast Conference, 2007
2) Beyond Green High Performance Building Award – Sustainable Building Industry Council, 2007
3) Commercial Project – Pacific Coast Builders Conference, 2007
4) Gold Nugget Grand Award for Excellence and Value for Best Sustainable Commercial Project – Pacific Coast Builder Conference, 2007

5) Honor Award for Excellence in Design American Institute of Architects – Pasadena Foothill Council, 2007
6) Honor Award – United States Green Building Council, 2008
7) LEED Platinum Certification – United States Green Building Council, 2008
8) American Architecture Award – Chicago Athenaeum, 2008
9) Honor Award for Excellence in Design – American Institute of Architects – Los Angeles, 2008
10) Merit Award for Savings by Design – American Institute of Architects – California Council, 2008
11) Green Good Design Award – Chicago Athenaeum, 2009
12) Honor Award for Excellence in Design – American Institute of Architects – California Council, 2009
13) Iconic Awards, Public Architecture with Distinction – German Design Council, 2016

The project is the first museum in the world to receive LEED Platinum Certification.

Umwelt Arena – Interactive Museum for Sustainability & the Environment Zurich, Switzerland

Courtesy to to Swiss Solar Solutions on Umwelt Arena

The place is around half an hour from the Zurich city center. The prime objective of the Umwelt Arena is to showcase sustainability in energy, transportation, and housing. It is an interactive museum of this type where one can have ideas on energy and environment concerns. More than 100 companies and organizations participated in demonstrating real experiences in renewable energy, sustainability, and interaction with nature, as well as how it will impact our lives in energy and mobility. There are forty-five exhibits illustrating how one can have personal comfort within our budget on energy and mobility.

In Switzerland, the buildings consume around 40 percent of the energy. However, around 70 percent of the buildings constructed before 1980 do not meet the required standards. Building designs should be well thought out. Since then, building codes have called for a heavier insulation and construction standard

The arena is designed for all ages to learn, especially for the young generation to learn and plan the future of the world they are inheriting. It is really worthwhile to pay a visit to the museum while traveling in Europe.

Topics of the Exhibits:

1. HVAC (Heating, ventilation, air conditioning) controlling system
2. Lighting
3. Gardening plants
4. Building materials
5. Building envelope and technology: a) color solar facades; b) solar railings
6. Sustainable construction
7. Battery storage
8. Water and management, especially waste water
9. Energy and mobility
10. Smart energy lab
11. Energy network of the future
12. Innovative building automation

In light of all the activities happening at the arena, it only consumes around 50 percent of the energy it generates annually, which would allow it to take care of certain emergency situations or help out neighbors when they are in trouble. We should consider doing it with 100 percent renewable in all industrialized countries especially in the US.

Owner: Umwelt Arena AG

Architect: Rene Schmid Architect AG

Awards:

1) Swiss Solar Prize 2012
2) European Solar Prize, 2012, in Berlin
3) Norman Foster Solar Award, 2012
4) Zurich Climate Prize, 2016

Solar system approx. 5,300 square meters, equivalent to the size of about 20 tennis courts. Solar thermal plays a significant role as well.

Annual energy production: 540,000 kwh, maximum output at 750 kWp. At this amount of energy generation, it could save over 100,000–150,000 liters of fuel oil on heating in one year.

Sunflower at the Umwelt Arena operates according to the Sun's trajectory

Chapter 3

NET ZERO ENERGY AND MAGNIFICENT HOMES

Atlantis's first net-zero energy system – Microgrid in Montana

Courtesy of Lee Tavenner of Solar Plexus

The Sunslate system provides the perfect roof for a high-elevation, high-end home in snow country—the Montana Barn. In 1999, Atlantis did not yet have the thermal system beneath the PV system, but was able to come up with a microgrid stand-alone net-zero energy by just installing the Sunslate. The initial plan was to install 7.4 kw of Sunslate, but the customer later wanted to have enough energy to support two more houses that are around 150 feet away, which resulted in adding Sunslates to a total of 14 kw.

The Sunslates charge twenty-four two-volt batteries so that the pump can bring water from the 450-foot well to one of two 1,500-gallon tanks. The filtered waste water will go to the other tank, which is used for plants and farming. The tank that has the clean water from the well is used for domestic hot water as well as warming the other two houses. After twenty-plus years of operation, the system is still in operation. Above all, the contractor said that the Sunslate is still the best product in the whole system after all these years.

The whole system and operation

1. The 14 kw Sunslate system charged up the twenty-four units of the 2-volt battery. There are two 1,500-gallon tanks in the barn. One contains fresh water pumped up from the 450 foot well.
2. The waste water from the houses goes to a nearby lagoon; around it, are marshlands The water is being filtered and prior being pumped back to the second tank, where it will be purified. The water will be for farming nearby.
3. The balance of the electrical energy will support the domestic hot water, warming the house and all the lighting requirements.
4. This is a totally self-sufficient net-zero energy installation, able not only to take care of the energy need of the three buildings. it can also handle supplying water for farming.

At a remote area like this one, connecting to the grid is quite a challenge. Apart from great concern, especially in the winter months in Montana when there are heavy snowstorms, the chances of power outages are very high. One could be trapped in a very dangerous position for a great number of days. Solar roof tile does not have any moving parts, so the likelihood of having problems is quite small. The only moving parts are the pumps. On a positive note, we are happy to learn that after all these years, the Sunslate on the Montana project has performed well despite going through many adverse weather events.

Energy Generation Systems – BIPV and BIPVT systems applications

BIPV and BIPVT are basically solar modules that are part of the weather skin of a home or a building. Sunslate is a solar roofing tile. The above examples have demonstrated the 14 kw system can make all three buildings totally independent on energy. In light of all the latest advancements in solar technologies, it is getting easier to install building-integrated solar products. Apart from being more affordable and aesthetically more pleasing, it also has a better investment rate of return especially on the BIPVT even though the initial cost is a bit on the high side. One may ask, "How so?" The current prices of standard modules are on the order of around 18–19 percent solar energy conversion efficiency. However, in the case of Sunslate together with the thermal, it will probably range from 35 percent to over 60 percent conversion efficiency, subject to the location of the installation.

Summer 2007

The test results from New Hampshire: Two 1 kw systems setup side by side

1. 1 kw of Sunslate
2. 1 kw of Sunslate with thermal (BITERS)

3 weeks in Brentwood, New Hampshire

Total sum of degrees	665.9
Volume	120 gallon tank
Density	8.33 pounds/gallon
Energy	665,633.6 BTU
	3,410 BTU/kwh
Power	195.15 kwh
Total Power generated by Sunslate	54.6 kwh
Total Power generated by Thermal	195.15 kwh
Total Power by the BITERS	249.75
Approx.	250 kwh
Temperature range:	55–102 degrees F

Apart from the above results, as the temperature reach 85 degrees F, the conversion efficiency of the Sunslate without the thermal system started to degrade, and as the temperature exceeds 100 degrees F, it reached around 10 percent degradation. The thermal energy extracted from the tubing could also reduce air-conditioning cost in the summer.

Based on these results, we decided to develop the BITERS system.

Though the experiment only covered a three-week period, we believe a sizable part of the country has this range of temperature. In addition, recent advancements on the solar cell technologies have reached the order of 21 percent conversion efficiency. Therefore, the module conversion efficiencies would be in the order of 18 percent. Once the thermal system is included, it would easily reach or go above 50 percent conversion efficiency in a place like California, which could make the house or building totally self-sustaining or net-zero energy. In case, some houses have swimming pools or whirlpools that can use the excess energy to extend use by a couple more months (1).

Looking at solar panels today, they offer around 18 percent solar energy conversion efficiency. In 2007–2009, efficiency was around 11–12 percent. Based on this assumption, the electrical energy generated would be 50 percent higher. At 54 kwh in 2007, it will be at 81 kwh on a 1 kw Sunslate system in 2022. Including the thermal will be 276.15 kwh on 1 kw Sunslate, which will be 3.4 times that of the 1kw Sunslate without thermal. At 18 percent solar conversion efficiency, it will be at 61 percent solar conversion efficiency with thermal.

Based on the New Hampshire experiment on the BITERS, performance of the thermal system that demonstrated the solar conversion efficiency is quite impressive, aside from the cooling effect on the PV cells, which makes a difference. Starting at 85 degrees F, the conversion efficiency began to degrade, reaching 10 percent at over 100 degrees. In addition, the cooling effect on the solar cell would also increase the longevity of the solar cells, especially in extreme weather conditions where the temperature range throughout the year varies from 15 to 100 degrees F. The stress on the cells could create cracks and eventually start to degrade the conversion efficiency and solar degradation (2). As mentioned, the ability to maintain the performance of the solar product for a longer period of time not only increases the return on investment, it means greater sustainability.

In light of the solar conversion efficiency of the BITERS system, say in a place like California, an average home does not require a sizable solar system to make it totally self-sustaining. A pool would be quite helpful because it could be used as a heat sink in the daytime, and at night, warmth could be extracted for the home through a heat pump. However, in New York where most homes do not have pools, geothermal could be used, or instead a large tank that could be bury in the ground is also a possibility. The heat collected in the daytime could be stored in the tank and extracted to warm the home at night and during cool days. Geothermal sometimes could be a problem where the buried tubes are not operating optimally.

Sunslate BITERS after a snowstorm

The Sunslate BITERS installation on the Wisconsin Rapids net-zero energy home. In addition to the thermal system, it also has the geothermal system.

Courtesy to Dr. Todd Duellman

It is a 2,500-square-foot house plus a 900-square-foot basement. The Sunslate system size: 9.5 kw with thermal system beneath the solar roof tile. Together with a geothermal system and well-insulated walls and ceiling, it managed to achieve net-zero energy. In addition, during winter months, the pavement by the garage melts the snow as well, so the owner does not need to shovel it because the geothermal system is right beneath it. On top of that, as one can see, the snow has slid down the Sunslate side of the roof. The house was built around 2009, and the solar installation was completed around 2011. In the same year, the house was awarded a Gold LEED (Leadership in Energy and Environmental Design) rating.

Sullivan Estate

Fig 3: Sunslate BITERS at the Sullivan Estate with Solar Tree in Hawaii

The 5.2-acre estate was built in 1962 by Maurice Sullivan, who was the head of the Foodland Supermarket dynasty. When Dr. Juergen and Karin Klein came to the island in 2004 from Switzerland, they were looking for a place to set up a spa, and once they found a perfect location, they bought the property. They began renovation in 2005 and completed the process in 2008. During the renovation, they installed the state-of-the-art Sunslate BITERS solar roofing system that Atlantis had just developed. Considering the moderate climate in Hawaii, the property was able to achieve net zero energy and was believed to be the first spa in the US to achieve it at the time. However, it would also be subject to how energy is managed on the property. In addition, it was turned into an exclusive spa retreat resort for celebrities or anyone who would like to have privacy and complete satisfactory services.

Solar system size:
Roof size: 9,250 square feet
Sunslate: 3,700 pieces, approx. = 48–50 kw.
Bare slates: 4220 Eternit
Starter slates: 205

The electrical energy generated by the Sunslate goes through the inverter and feeds to the local grid to offset the energy being consumed. On the installation of thermal systems, the BITERS beneath the Sunslate covers a sizable portion of the roof. The thermal energy is useful for domestic hot water including a Jacuzzi, hot baths, a warm seawater pool, and kitchen appliances. The temperature fluctuation in Hawaii hovers between 78 and about 90 degrees F. Looking at this temperature range, the solar conversion efficiency could be in the order of around 50 percent, which is quite high. On the location, there is also a solar tree which pumps the water on the fountain. As the sun's intensity goes high, so does the water.

Since the JK7 Spa opened its doors in 2008, it is one of the top spas and skin care medication facilities in the world. However, it is very private, and that is why, one does not see many ads about it.

Sunslate BITERS – Cherokee Home in Raleigh, North Carolina

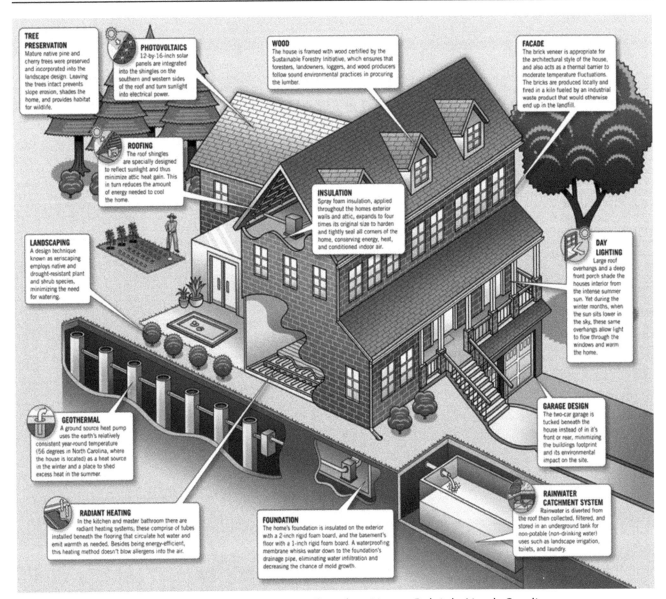

TREE PRESERVATION
Mature native pine and cherry trees were preserved and incorporated into the landscape design. Leaving the trees intact prevents slope erosion, shades the home, and provides habitat for wildlife.

PHOTOVOLTAICS
12-by-16-inch solar panels are integrated into the shingles on the southern and western sides of the roof and turn sunlight into electrical power.

WOOD
The house is framed with wood certified by the Sustainable Forestry Initiative, which ensures that foresters, landowners, loggers, and wood producers follow sound environmental practices in procuring the lumber.

FACADE
The brick veneer is appropriate for the architectural style of the house, and also acts as a thermal barrier to moderate temperature fluctuations. The bricks are produced locally and fired in a kiln fueled by an industrial waste product that would otherwise end up in the landfill.

ROOFING
The roof shingles are specially designed to reflect sunlight and thus minimize attic heat gain. This in turn reduces the amount of energy needed to cool the home.

INSULATION
Spray foam insulation, applied throughout the homes exterior walls and attic, expands to four times its original size to harden and tightly seal all corners of the home, conserving energy, heat, and conditioned indoor air.

LANDSCAPING
A design technique known as xeriscaping employs native and drought-resistant plant and shrub species, minimizing the need for watering.

DAY LIGHTING
Large roof overhangs and a deep front porch shade the houses interior from the intense summer sun. Yet during the winter months, when the sun sits lower in the sky, these same overhangs allow light to flow through the windows and warm the home.

GEOTHERMAL
A ground source heat pump uses the earth's relatively consistent year-round temperature (56 degrees in North Carolina, where the house is located) as a heat source in the winter and a place to shed excess heat in the summer.

GARAGE DESIGN
The two-car garage is tucked beneath the house instead of in it's front or rear, minimizing the buildings footprint and its environmental impact on the site.

RADIANT HEATING
In the kitchen and master bathroom there are radiant heating systems; these comprise of tubes installed beneath the flooring that circulate hot water and emit warmth as needed. Besides being energy-efficient, this heating method doesn't blow allergens into the air.

FOUNDATION
The home's foundation is insulated on the exterior with a 2-inch rigid foam board, and the basement's floor with a 1-inch rigid foam board. A waterproofing membrane whisks water down to the foundation's drainage pipe, eliminating water infiltration and decreasing the chance of mold growth.

RAINWATER CATCHMENT SYSTEM
Rainwater is diverted from the roof then collected, filtered, and stored in an underground tank for non-potable (non-drinking water) uses such as landscape irrigation, toilets, and laundry.

Fig 1: BITERS installed in Green Cherokee Home, Raleigh, North Carolina

This project is the initial serious trial of the BITERS (Building Integrated Thermal Electric Roofing System), the solar hybrid system which could convert the sun's energy both into thermal energy and electricity. It is totally building-integrated. The Sunslate, which is a BIPV product, converts solar energy into DC power, and the metal batten together with the tubing installed beneath converts solar energy into thermal energy.

It is basically a traditional red brick home, but it has all the latest energy generation and energy-saving technology products and systems.

1. The Sunslate: Solar roof tile which generates electrical energy which reduces the electricity consumption of the building. Beneath it is the solar thermal system which collects the heat that assists in domestic hot water as well as cooling off the space, resulting in air-conditioning cost savings.
2. Radiant floor heat: Located at the bottom of the house as shown. Rainwater is collected from the roof and goes the tank shown at the bottom of the building. As we all well know, water is a precious commodity.

Together, the design of this house has these items in mind:

a. On the environment side, it has the ability to recycle 90 percent of organic waste.
b. Consume 50 percent less water.
c. Retain 95 percent of stormwater collected and reuse it.
d. Recycle 75 percent of all demolition and construction waste.
e. Provide great indoor quality of air which includes that 95 percent of all the products that contains little or no VOC (volatile organic compounds).
f. On sustainability areas: 1) Energy Star Rated windows and appliances, 2) the best type of insulators, 3) a ground source heat pump, and 4) a solar thermal water heater.
g. It also creates wildlife habitat.

Awards:

1) US Department of Energy Gold Award
2) National Home Builder Gold Award
3) Highest Honor – National Association of Home Builders Research Center
4) City of Raleigh overall environmental award (all categories)

5) LEED platinum award (first in southeast US)
6) National Wildlife Federation: Certified Wildlife Habitat for backyard
7) Gold certified under the North Carolina Solar Center HealthyBuilt Homes program
8) First home in the nation built in typical subdivision under NAHB Model Green Home Guidelines
9) International Builders Show (IBS) Gold Energy Value Honoring Award (EVHA), 2008
10) NREL (National Renewable Energy Lab) Gold Award for successful integrating cutting-edge, energy-efficient features
11) Designed by William McDonough – World Renowned Solar Technology Architect

This project, done in 2006 in Raleigh, North Carolina, demonstrated that the solar roof tile together with the thermal system beneath could make substantial improvements on the solar conversion efficiency, which could enable the house to achieve net zero energy. Considering recent advancements in solar cell technology as well as storage and generation technologies, the net zero energy house is now a reality and is going to be affordable to average households in the country. The Cherokee Home added a lot of additional features.

To get the system started, one can just have the solar electric and thermal together with the storage and generation system to make the house net zero energy. All the other bells and whistles can be installed later.

It is the first BITERS installation, which happened in 2006, that managed to capture a substantial amount of solar, with conversion efficiencies to above 30–40 percent in an area like North Carolina; efficiency could be even higher in Georgia or Florida. Looking at the performance, Atlantis decided to set up a test site in the summer of 2007 in New Hampshire, which demonstrated that the BITERS at temperatures ranging from 55 to 102 degrees F, the solar conversion efficiency was around 50 percent. Back then, the solar cell efficiency range was 12–14 percent. Today, the range is 20–22 percent, which would yield on the order of 60 percent efficiency on the combined electric and thermal systems. Places such as Southern California and Arizona could easily make buildings or houses net zero energy by applying the Sunslate BITERS.

The Sunslates would be connected in series horizontally across the roof through our custom designed j-box. The series connection of the Sunslates would form a string, and the number of connections in series would be subject to the design. The Sunslate thermal and electric system would consist of the Sunslate mounted on a metal batten which also holds the thermal tubing. Sunslate is the photovoltaic solar roof tile that converts solar energy into DC electricity. The metal batten holds the thermal tubing which collects solar energy and converts it into thermal energy through the liquid running through the thermal tubing.

The thermal part of the system should be installed first. The metal batten will be mounted horizontally across the roof onto the battens that run vertically up the roof. Once the metal battens are installed, the thermal tubing can be mounted on the gutter of the batten horizontally across as well. The spacing between the metal battens could vary, subject to the size of the Sunslate and the thermal required specifications. As the metal battens are more tightly spaced, more thermal tubing could be installed, resulting in higher thermal energy conversion efficiency. The thermal tubing is connected to the pump and a tank with a heat exchanger. The metal batten has three applications: 1) mounts the thermal tubing, 2) transmits the heat, and 3) mounts the Sunslates.

The metal battens are used to mount the Sunslates. Hooks are used to secure the mounting of the Sunslate. One hook holds down the Sunslate at the top, and the bottom hook supports it as it overlaps the adjacent Sunslate. Holes are drilled in advance according to specified positions on the metal batten, which saves time and simplifies the installation. The specified position would be based on the position of a potential micro-inverter, or numerous strings could be connected in parallel onto a regular inverter or numerous Sunslates could be connected in series feeding to a regular inverter, subject to the size of the inverter and the system specifications. The design would depend on the specifications of the system. The inverter could convert the DC electricity to AC and feed into the grid. However, if one would like to have a stand-alone system, one could use the Sunslate system to charge up batteries by configuring the Sunslate at 12, 24, and 48-volt strings or a hydrogen fuel cell system (7).

The customer on this project put a new patio extension on his existing home and uses the Sunslate BITERS. It is a relatively small system at 2.7 kw, and he has realized substantial savings on his utility bill. Including the new extension, the building is around four thousand square feet, and he is a heavy consumer of electricity, with a Jacuzzi as well as an outdoor swimming pool, which he is able to use for an extra month or two. His monthly electric bill is well over five hundred dollars, and the Sunslate BITERS cut it by almost half. Later on, the customer might have added more Sunslates to further increase the power generation and made it net zero energy. Consequently, he has been a very happy customer after the installation.

Virginia is considered as a mid-Atlantic state where climate is somewhat milder than the Northeast. Just imagine by looking at average American homes, they are around 1,500–2,500 square feet. A 2.5–3 kw BITERS system could make the home net zero energy in most mid-Atlantic states, including Maryland, Pennsylvania, and Washington, DC. However, it can get quite cold in the winter months which would be a little below freezing. Similar places in Europe would be London, Paris, and Geneva. BITERS system applications, in addition to roof tops, could be installed in façades and curtain walls together with Sunslate.

The Sunslate BITERS could be ideal in a relatively tropical or Mediterranean climate, in places like southern France, Italy, Greece, or Florida, the Bahamas, and Singapore, to achieve total self-sustaining. In all these locations, winters are relatively mild, if they have winter at all. In situations when temperature goes below 30 degrees F, a reasonably good-sized storage tank would be sufficient to maintain the thermal energy throughout the winter months. One customer installed a Sunslate BITERS in northern California; throughout the last couple of years, he managed to store more energy than he needed and released some energy on other applications. Considering the current heat wave in Europe and the US, the BITERS could cool off indoor temperatures and store energy for winter months

An 83-year-old house turned into
Beauty and the Beach
Eighty-three-year-old Beach House in 2011, front

Beach House facing Truesdale Lake in 2011
Beach house in Truesdale Lake,
South Salem, New York

The house was built in 1932 overlooking Truesdale Lake, Westchester County, NY. In late 2011, they started renovation and came up with a design to provide maximum comfort as well as minimal carbon footprint that was completed in 2014 by architect and builder Sylvain Cote. Optimizing the solar energy performance on the tight south facing roof, the Atlantis Energy Systems electric roof system was recommended, the Sunslate. In addition, the Sunslate satisfied the historic and zoning code (reference).

Installed components:

1) A 3.30 kw Sunslate PV provides electricity and domestic hot water. Unico high velocity small duct system that cool effectively in removing humidity and delivers conditioned air throughout the walls and ceiling with hardly any energy loss.
2) On heating, it has Viega system radiant heat which is very healthy.
   In addition, there is a Fantech energy recovery ventilator that constantly brings the fresh air from the outside which improves the indoor air quality. All systems can be controlled remotely to optimize energy efficiency.

Awards:

1) Best Renovation Home of the Year by Green Builder, 2015
2) LEED Platinum Certification of 90

3) Energy Star Certified at HERS (Home Energy Rating System) rating 30

Ref: Matt Power: Editor-in-Chief, Green Builder Magazine, January 2016

ConclusionLooking at all the installations shown above, BITERS and Sunslate is one of the major solutions to bring houses and buildings to net zero energy. In some locations where the design of the building requires additional system support and in locations like California, southeastern states could probably does not require. The initial cost could be on the high side, but in the long run, it could be lower as well as more energy-independent.

Magnificent Homes

Swiss Magnificent Home
Courtesy of Swiss Solar Solutions

Magnificent Home
Custom Glass/Glass Modules

All three magnificent homes shown above are still with solar roof tile modules. The resident house in Tiburon, California, has glass/glass solar modules. The image is not shown. At the current moment, they are not a net zero home all. It will not be too difficult for them to achieve net zero, especially as they are located in California. If they would install a thermal system beneath the solar roofing modules, they would able to make it to net zero.

Magnificent Home, Langenthal, Switzerland
Swiss Megaslate installation

Location: Langenthal, Switzerland

System Size: 22kw

As to the residence in Switzerland, the system size is enough to achieve net zero energy status by using heat pumps. They applied three modulating heat pumps which start operation at zero amps, which can optimize energy with less stress on the systems.

The residence in Gloucester, Massachusetts, has had the BITERS system installed for more than twelve years and the customers have been very pleased on the performance of the whole system. Though it is not quite net zero, the system managed substantial reduction of the energy cost throughout the years. In the winter months, heat extracted from the BITERS substantially reduced the cost of heating cost in the building. The owner could install a geothermal system together with a heat pump, which could make it a net zero energy system; a large water tank together with a heat pump could achieve the same result.

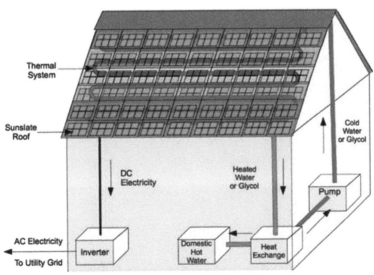

Reference:

1. BITERS, US patent no. Pao & Gottlieb 8,196,369 B2, Pao 8,201,382 B1, Pao 8,365,500 B2, Pao 10,958,212 B2, Pao 10,958,212 B2 plus numerous patents pending status.
2. https: novergysolar.com/understanding-the-degradation-phenomenon-in-solar-panels

Chapter 4

SOLAR APPLICATIONS ON TRANSPORTATION

"Dream today, reality tomorrow" Claire Marchand, Managing Editor, IEC, Geneva, Switzerland (1).

Japan started the maglev train back in April 2015 which broke the world record at 603 km/h. They anticipate to connect southern part of Shimagawa, south of Tokyo and Nagoya in center part of Japan by 2027. Train like the maglev train can be helpful to long distance commuters. Meanwhile, Elon Musk has thought of a vacuum transportation that would go even much faster, potentially at supersonic range.

Currently, in most major cities around the world, traffic congestion is a very serious issue. Local municipalities throughout the years have been trying all type of ways to resolve the problems, and yet they have not managed to do so. Over centuries, people have started two-dimensional dwellings, but to this day, the transportations systems are still basically one-dimensional. The subways to some extent are two-dimensional, but it is not totally the case because it is too expensive. Today, some major cities in the world have started using the cable cars to help out on the congestion in some of the routes. One of the main reasons that the airplanes are so safe because it is a three-dimensional travel. During takeoff and landing, they go from one-dimensional to three-dimensional. The tricky part is during landing, because then planes are going from three-dimensional to one-dimensional. A lot of time, they have to be on holding patterns for a while before they can land.

In addition, weather and other unanticipated conditions on the ground could affect the travel congestion. However, in light of all these potential concerns, to smooth out congestion, start looking into two-dimensional on ground operation to diversify and optimize our transportation capabilities in order to make a better flow of traffic conditions, which could be vital to our future.

Although, in recent years, a new generation of youngsters, partly due to increases in the cost of automobiles, would prefer not to own one, as the world's GDP goes, the ownership of private cars will continue to grow, probably at a slower pace. Motorists will continue to waste lots of time and money on traffic congestion, especially in large urban areas where the growth of demand surpasses capacity, especially in the case of single occupancy vehicles. These problems draw a lot of researchers' attention, and how these situations are resolved could affect economic growth and national security, especially when fuel prices are volatile and emission concerns due to climate change are high.

Throughout the years, a number of efforts have been made to improve the efficiency of traffic flow. Despite all the efforts, it has not been satisfactory. In the case of mass transit, increasing capacity during rush hours in high density metropolitan areas could affect the initial and last mile situations. Meanwhile, the rigid stop-and-go that forces passengers to wait for the unnecessary stops for every individual passenger creates low efficiency in reaching the destinations. Start and stop also increases energy cost. In case of low ridership, it affects the cost as well. In the case of internet data packaging and communications, if some small number of passengers can be grouped to small, light vehicles and allow them to move automatically through a network which is like a physical internet, the approach could be applied in transportation to mitigate congestion.

The concept of PRT (Personal Rapid Transit) is traced back into the 1950s when Donn Fichter, a city transportation planner in New York City, began researching it as an alternative transportation solution which would be automatic. Proposals were made in considering medium to low population density areas. Though the automation concept of the PRT was established at that time already, and throughout the years advancements have been made, there was hardly much attention paid until a bill was introduced but never made it through the House of Representatives. However, after President Kennedy came into office the bill was reintroduced, passed, and signed into law on June 30, 1961. He guided the Congress to look into transit situations and later stated, "To conserve and enhance values in existing urban areas is essential. But just as important are steps to promote economic and livability in areas of future development. Our national welfare requires the provision of good urban transportation, with the properly balanced use of private vehicles and modern transport to help shape and serve urban growth" (Wikipedia).

Looking into the development of automated urban transportation systems, the objective would be to achieve deployment of the PRT through a highly interactive process between a variety of professionals and politicians, including private citizens. People can appreciate that PRT systems could be a complement to private cars as well as providing more convenience, which could arouse more public interest as a dream to pursue, including the potential of investments of time and resources.

Before going any further, let us look at the history of PRT. It initially got interest in the late 1960s when HUD (Department of Housing and Urban Development) of the federal government looked into studies. The DOT (Department of Transportation) was not around until 1970. Consequently, HUD set up UMTA (Urban Mass Transportation Administration) as part of the organization. Of all the study reports they did, two of them attracted their attention. 1) The Stanford Research Institute developed various new concepts in moving sidewalks to PRT to dual mode and to estimate their economic benefits to the US. 2) In a study by the GRC (General Research Corporation) of Santa Barbara, the objective was to model alternative transport systems in actual cities so that they could determine how they would perform as compared to conventional transport systems. A selective group of seventeen specialists from different fields considered Boston as a relatively typical large transit-oriented city, with Houston as large and Hartford and Tucson as small auto-oriented cities. Viewing the study results through computer modeling, they highly recommended new transit systems due to potential population growth, which could seriously affect the transportation system when they continued to use conventional means of traffic operation by using automobiles. Deploying PRT would be the solution to resolve the traffic congestion.

It turns out that the GRC study has been most influential of all the HUD studies for two reasons. The article published by *Scientific American* summarized it quite clearly and became a classic situation where it provides solutions to traffic congestions issues by applying new technology. In addition, the report convinced GRC chairman Ben Alexander regarding the importance of creating a national commitment in developing new transportation technology as well. He successfully managed to testify before a congressional committee, which resulted in bringing the PRT and dual mode concepts to Washington.

There was an unfortunate change in administration when the HUD report was released. Consequently, the PRT development ended up playing a minor role in UMTA's agenda. Accordingly, there were concerns that implementing the PRT could affect the collapse of the existing transportation system in our nation. Instead, the administration decided to provide more support in improving the buses and railroad systems. Countries like Germany, France, England, Japan, and Sweden, including the US, made attempts to develop PRT technologies throughout the sixties and into the seventies.

England – Cabtrack in 1960s

The Cabtrack is a real PRT which was initiated by L. R. Blake, who worked in Brush Electric Company got to come to the US and looked into the Alden staRR car invented by William Alden, together with other automated transit systems. Around 1967 he wrote an article on a transportation concept relatively similar to the PRT which would be suitable for cities and towns in England, which he called Autotaxi. He later started a company and sold it to Brush Electric. Over there, the executives tried to persuade the minister of transportation to continue to take action on the concept and later cooperated with the National Research and Development Board on a 50 percent joint venture to develop the Cabtrack to implement a test track. They were looking at a budget of 250,000 pounds. Afterward, the Royal Aircraft Establishment at Farnborough, Hants, set up an urban transport group and renamed it Cabtrack (11). To get the project started, the initial phase was a nine-month study, and the report was due in December 1968. Afterward, the Royal Aircraft Establishment signed an eighteen-month contract and additional contracts that culminated into setting up a one-fifth-size scale model for testing. By March 1974, they issued the last report. It was a comprehensive system study of the PRT by the government. Not only did it provide substantial technical development information, but it also provided extensive demand and layout process analytically. On top of it, a large variety of system control schemes were included, together with the pods operation specs at a minimum spacing of six seconds. A contract was awarded to Robert Matthew, of Johnson-Marshall Partners, a large British architectural firm, to study how the Cabtrack system could work in certain part of London. The report of the study was published in the May 1971 issue of the *Architects' Journal*, which was the earliest series of studies regarding to the visual impact of the overhead guideways on the PRT. However, due to election results in England, the newly appointed minister of environment decided to terminate the Cabtrack program, which was quite unfortunate. It was supposed to be the earliest professionally developed high-capacity PRT program in the world. The last report is still of great value in methodology and performance results which could be very helpful if all these valuable data had been put together in a book that professionals could have referenced in later years.

Japan – CVS (Computer-Controlled Vehicle System)

The system was developed by the Japanese Industrial Consortium in the 1970s. The initial development of a PRT system consists of four-passenger cabins at two thousand pounds in weight. To better understand the PRT system, they decided to have a scale model of a thousand vehicles to simulate the operation. in a suburb of Tokyo, in 1972, they constructed a 4.8 km guideway consisting of sixty pods. In 1978, after constructing the second phase, the organization thought of opportunities to deploy the system while looking at a number of proposals to install it in Baltimore around the late seventies. In light of a series of studies, they decided to cancel the project because of the issues listed (3).

1) The guideway size, 3 meters in width and 1.8 meters in depth, presented serious concerns about cost and the visual aesthetic impact of the system.
2) Potential problems in serious weather conditions.
3) The rides were on the rough side.
4) Loading and unloading of passengers and the carriages' capacities.
5) Weight and operating costs.

Later on, they visited the United States to see the advancements of the PRT there as part of their study. Looking at the situation, they realized they had too many barriers to overcome as well as the market projections, and they decided not to continue the project.

Germany: Cabinentaxi

The German Ministry of Science and Technology was interested and aware in 1970 that Messerschmitt-Bölkow-Blohm and Demag were working independently on PRT projects with similarities to each other. Meanwhile, they were quite inspired by the work done by HUD in the US. So they were requested by the ministry to work together and share their developed knowledge, including their resources and finance support from the Bundesministerium für Forschung und Technologie (BMFT), which is the German Ministry of Research and Development.

They came up with a program in analyzing different suspensions, switching, motors, and guideways that led them to come up with different cabin sizes and ways to operate the cabins above and below the beam. In addition, the vehicles will be running on rubber wheels and operated by linear induction motors on both sides of the pod. This way, the vehicles can run close to each other, which could be on the order of seconds apart. After the study on the control system, they selected asynchronous controls instead of synchronous or quasi-synchronous because the asynchronous control system is more flexible to handle in case of need to adjust the speed or change stops of the vehicles. They tested the system on May 30, 1973, and later in October 1974; they managed to successfully demonstrate the system to the Minister of Science and Technology as well as the German press. Later on, they worked and tested the reliability and the human factors of the operation as well as maintenance, which resulted in arousing interest in deploying the system to some cities. They were considering installing the PRT system in Freiburg and Hagen, which they thought could be successful, and perhaps other cities as well (3).

In 1975, Raytheon Missile Systems Division looked into numerous PRT system development and had thoughts of licensing the Cabinentaxi to be operating in the US. After reviewing the situation, they realized that it was not their core business of operation and decided not to pursue it. At the same time, Demag and Messerschmitt-Bölkow-Blohm also was promoting the PRT system in other US areas, for example, Minnesota. There was a comprehensive elevation on the automated guideway at the central business district in Indianapolis. On top of that and around the same time, Germany was in the process of constructing a twelve-passenger pod system in Hamburg. In light of the economic conditions around the early eighties, the German government had to cut back on their budget and decided to terminate the project. This was very unfortunate because the PRT system deployment might have been able to solve the traffic conditions.

Here are some of the unique features and the advantages of the Cabinentaxi PRT system.

1. The cabins are relatively small with seats available at the stations and will be ready to take on passengers on demand.
2. The vehicle goes directly nonstop to its destination.
3. Operation speed would be around 30–40 kph.
4. Located on the elevated guideway, it is separate from other modes of transportation.
5. The operation is fully automated.
6. Using the linear induction propulsion system provides a better control on vehicles' separation in operation to be below ten seconds spacing.
7. Noise level is low, and there is no air pollution.

France: Aramis

Initially, the French system started with a four-seated PRT vehicle that uses rubber tires on the wheels. The vehicle operated on their unique type of variable reluctance motor. They started the operation through Gerard Bardet with a budget of ten thousand francs. A French aerospace company, Engine Matra, bought the patent and initiated the development of the PRT system. Later on, in the 1970s, Matra got the contract on Aramis from Datar, and in April 1973, they started full-scale testing at Orly International Airport in Paris. By summer 1974, they were able to do the first phase in checking out the basic concept of the system. Around the same time, Matra received a contract from the Paris Métro authority to show a demonstration to the public audience of the PRT system in a Paris suburb. The proof of concept on the initial phase was reliable and safe (Wikipedia).

The French Aramis was unique as compared to other PRT systems because the system was linked together in platoons instead of operating individually. However, the vehicles were separated by about thirty centimeters, but in case one of them had be separated from the platoon to go to a station, this could be handled by the in-vehicle switch. The vehicle could leave the station and catch up with the previous platoon. The type of arrangement demonstrated that the platoon could quickly increase the capacities and the ability to load and unload at the same time.

The design of the platoon concept is to handle the circumferential system around Paris. In this case, the system is not compatible to a network type of operation like the PRT. In the platoon system, it would be difficult to control the brakes in bad weather. So, to get the project going, they later on tried a ten-passenger vehicle which potentially might present security concerns. Consequently, it was eventually decided to cancel the program. However, they could have installed the other type of PRT in other cities—Marseilles, Bordeaux, and Lyon.

Sweden: Gothenburg

Gothenburg is the second largest city in Sweden, and their transport authority was quite inspired by the Cabtrack PRT system from England and took the lead in looking into it. However, the city had thoughts of a subway, but it would be too costly to build because the greater part of it involves solid rock. PRT is an attractive solution to be considered. In March 1973, the Gothenburg Transport Authority studied the international status of the PRT development and realized that none of them were quite ready yet to be implemented in Sweden. Instead they decided to expand their tram system while waiting to see how other countries made the advancements on PRT. Throughout the years, they have been observing the progress internationally, and later Sweden had renewed interest in the PRT system.

Therefore, in 1994, they decided to do a study and see how to implement the PRT system on an elevated guideway in an old town in Gavle with a population of ninety thousand. The report provided attractive graphic images of the PRT system around the city. They also provided the map layout of the network, including the design on a variety of support columns of the guideways. There were perspective drawings with color photos of the city in the report for three station locations that gave a three-dimension view of the PRT network. The study also provided the dimension of the structure, load bearing of the columns, and all the guideway elements including the environmental concerns that could be affected. Unfortunately, faced with the environmental concerns, the visual encroachment on the city's environment, and the visual intrusion of the stations, as well as the potential of noise that could affect the neighborhood, they decided not to build the PRT system.

Later on, Vectus, the Sweden PRT system, was being developed in cooperation with South Korea (3). In 2007, a full-scale system test track with that could be commercialized included all the major components was demonstrated by Vectus and approved by the Swedish Railway Agency. It was also opened to visitors. The Vectus PRT uses lightweight electric vehicles on the elevated track which is less costly than adding a new subway system or automobile highways. Consequently, Vectus in 2013 worked on a new guideway in Suncheon Bay, South Korea, which is a coastal town on wetland which would have minimal environmental impact concerns.

United States: The Morgantown PRT, West Virginia

The project started in the late sixties through the Aerospace Corporation set up by the Air Force. Around the period of 1968–1971, the Aerospace Corporation developed the whole PRT concept to an advanced level and proved the feasibility of a sizable system, on the order of thousands of vehicles, with spacing of less than one sixth of a second at the operating speed of sixty miles per hour. The organization studied the economic and the financial support interests on the PRT in Los Angeles, California, and Tucson, Arizona, and presented substantially on the PRT advantages. Dr. Lawrence Goldmuntz at the White House OST (Office of Science and Technology), as Director of Civilian Technology, expressed great interest in the proposal to develop the PRT project with the Aerospace Corporation, which resulted in an announcement in January 1972. At that time UMTA was requested to allocate $20 million to support the development the high-capacity PRT system. In autumn 1972, the Department of Transportation approved the project and requested to work with NASA.

Unfortunately, the delays were due to political issues in Washington around 1973–1974. Later on, there was a change in administration as well as lobbyists opposed to the project, which brought it to an end.

On the positive side, concurrently, in the state level, Morgantown, West Virginia back in the late sixties, Prof. Samy Elias, head of industrial engineering at the University of West Virginia, initially looked at the PRT system and began to realize that there were some test tracks in the country. Consequently, he believed that it could be beneficially deployed at the university so that students and faculty members could travel around campus buildings—without delays and economically. It was very fortunate that they got the support of the UMTA along with congressional delegates. Together with the city of Morgantown and the university, political pressure from the state of West Virginia gave a great opportunity to get the project going. It happened that John Volpe was the newly appointed Secretary of Transportation who followed up on the proposal, and NASA's Jet Propulsion Lab in Pasadena was the project manager that signed the agreement in 1970 with an industrial group including:

1. The manufacture of the cabin vehicle was assigned to Boeing of Seattle.
2. The control system was assigned to Bendix Company at Ann Arbor, Michigan.
3. The guideway, station, and landscape design of the area was assigned to F. R. Bendix Engineering Company, Stamford, Connecticut.

These companies had never had any prior experience in handling or construction of these systems or even operating them, resulting in sizable cost overruns that were quite disappointing to the Congress and oversea visitors. Regardless of what happened, the Morgantown PRT system has operated for more than forty years accident-free except for one collision in 2016 due to computer failure on the cabin. It is an outstanding track record as compared to automobiles. In addition, the system is operating in great condition with relatively low maintenance cost (11,12). It is a great chance to evaluate the effectiveness of performance to meet the demand of getting the student and local residents to different parts of town. In prior years, the main transportation in the area was through automobiles. Traffic congestion was a serious problem, even in a small municipality like Morgantown, which has proven that the PRT is an alternative means of transportation to meet the demand of students and other passenger schedules (6). Meanwhile, it is time to proceed on two-dimensional transportation systems, especially considering the rapid growth of population in our urban areas. For some, it could be a pleasure trip around the city.

As most of us know, train switching is controlled by the tracks, but in the Morgantown PRT, switching is controlled from the vehicles, which provides a much more versatile operation. Although the Morgantown PRT system budget was substantially overrun, in light of how well it has performed and innovations that were developed and well implemented to commercial standard, the whole project was well worthwhile. The PRT handles around eighteen thousand passengers per day, which could be around three million annually.

As time has passed, there were some minor maintenance problems because some parts, especially on the electronic components side, were no longer available. Therefore, the maintenance had to redesign some parts of the system based on the original principles. However, some advancements were made throughout those years. The system has also set a standard that provides a reference for the current generation. Over recent years, quite a few scientists and engineers visited the site. Though there are other similar PRTs in other parts of the world, none of them to this day is comparable to the Morgantown operation in terms of size and sophistication, especially as it was commissioned in 1975, almost half a century ago.

The Ultra PRT system at Heathrow Airport in London only goes between Terminal 5 and the business parking lot, which is only couple of miles away; this was commissioned around six years ago. The Masdar City in Abu Dhabi of UAE also uses the Ultra PRT system; the Vectus PRT system is used in Sweden and South Korea and Park Shuttle in the Netherlands. Consequently, the Morgantown PRT system is a model that can be looked at to see how new advancements could be built on today's control, electronics, and software technologies.

Morgantown PRT Operation

The system covers a guideway length of around nine miles and five stations in addition to two maintenance locations. The city is built along the river valley, and the PRT transportation system was built according to landscape along the Monongahela River. The guideway of the PRT meanders along the riverbanks in the older part of town, between the Walnut and Beechurst train stations. The northern part of the track curves back to the northeast through a small valley. Initially, there was limited space, so the town and the college decided to develop and expand the space. The Morgantown PRT has seventy-one cabins with rubber tires. Each one has eight seats but can hold a total of twenty-one passengers. All of them managed to handle around 1.5 million miles each year.

For the most part the guideways are elevated, and it goes up to 350 feet around the Walnut station and the medical center. In order to allow the vehicle to go directly to a destination without stopping, there are bypass lanes and turnaround channels in the middle three stations that provide options for the pods. It was a significant advancement back in the seventies and a great concept that makes the PRT stand out. As in the case of trains and buses in a linear operation, there are no bypass lanes to make stops.

As to system operation, there are three modes.

1. Demand mode
2. Schedule mode
3. Circulation mode

During peak periods, the demand and schedule mode will operate, while the circulation mode will operate at off-peak times. The demand mode will obtain the on-demand request of the system to optimize the number of required vehicles at peak or limit the number of required vehicles in order to save cost where the other two functions are being operated by the other two modes. However, in demand mode the system operates dynamically according to the passengers' requests and is controlled by the software which checks the passengers wait time and the actual capacities (4).

PRT Communications and Operations

The operation of the communication system handles the (VCCS) Vehicle Control and Communications Subsystem, which is part of the automatic software and hardware system installed on all the vehicles. It controls the vehicle movements and operations by interfacing between stations through handling the control signal to and from stations via the (SCCS) Station Control and Communications Subsystems so that they all know the positions of the vehicles at the guideway. Through an inductive communication link on the guideway control and communication link, the system sends interexchange information between the VCCS and SCCS, transmitting important information through tones and less important information digitally. The VCCS has the components listed below.

1) Embedded in the guideway are two antenna systems which provide two-way communication arrangements. One is a dual antenna assembly for sending, and the other receives information from the vehicles in a low-frequency electromagnetic signal.

2) On the vehicles, all the antennas are physically installed. As to the communication system, the antennas receive low-frequency signals. Then the information goes into the data handling system and then to the antenna of the guideway.

3) The data handling system receives the signals from the communication systems and decodes it into logical data which will instruct the system to send appropriate logical commands to the vehicles.

4) The control system will respond to the data provided from the data handling to control the operation of the vehicles such as brakes, steering, doors, and propulsion.

5) The support system synchronizes the data on the system as to the electrical energy conditions, electronic circuitry as well as the interface between the receiver and the transmitter.

The Vehicle Management System of Vehicles

The stations all have their own computer-controlled systems on the movement of the vehicles, together with protocols from the control centers that handle the routing on the incoming vehicles on loading and unloading of passengers, so that it can keep the track on. These data include

1. The vehicle and channel availabilities.
2. Open space at the berth.

The station system directs the vehicle to the correct channel through the routing logic decision system. Therefore, the speed and the space available are vital data as the vehicle enters and leaves the station. Switching to the appropriate channel, the timing control operation together with the turning control of the vehicle through the verification commands system are critical operations of the system. All the detail of positioning and timing of the vehicles has to be precisely controlled.

As soon as the switching operation is cleared, the vehicle gradually slows to a complete stop. The automatic stop is controlled of the PRT system within a range of six inches. To achieve this operation, the system on the vehicle interacts with the guideway stopping loop, together with the computer at the station, which signals the stopping loop at the channel in a precise response.

Once the vehicle has stopped completely, the doors open and the passengers are let out. So the door automatically closes and moves to the forward position according to the instructions programmed. Based on those instructions, the vehicle accordingly either stays, waits, or goes to the next station. In the event that are additional passenger requests at the current station, allocation of more vehicles to the customers' destination may be required. If no vehicle is available, then they will check at the other station to obtain vehicles.

When all the riders are on the vehicle, the door automatically closes, and the vehicle moves ahead. In case the door does not close, there are sensors that verify all is clear prior to moving on. Until then, the vehicle communicates with the station, which tells the central control system the vehicle's destination. As soon as the time to destination is determined, detailed information regarding the positioning of the vehicle at the station is be provided, which minimizes the likelihood of a collision. As the vehicle starts moving, instructions are sent to the vehicle by the platform to accelerate up the ramp while moving toward the guideway.

Guideway and Vehicle Operations

As the vehicle started going toward the guideway, the station system monitors through the computer that controls the data listed below.

1. Vehicle ID
2. Vehicle location at all time
3. Vehicle destination and switch condition
4. Vehicle speed
5. Door condition
6. Vehicle braking condition
7. Vehicle operation concerns

All the exact data on the information shown above will be communicated between stations by the guideway detector through the main control. The main control system also provides data to the guideway and the destination station, so that they are prepared to receive the vehicle. Any problems or system malfunctions are being monitored in order to prevent unforeseen mishaps. Computer programs are able to detect when the vehicle comes close to the station in order to allocate the right space at the berth. However, in case there is no space, the vehicle is required to wait at the ramp while space is being made available.

Safety – Personnel and Passengers

The Morgantown PRT system, aside from developing a very sophisticated automatic transportation system, also made substantial efforts on safety as well. They identify potential situations hazardous to passengers and personnel and eliminate them during design and development. During test runs they also look into components that could raise safety concerns. They also consider human factors and how to prevent mishaps from occurring. This is how the PRT system managed to achieve seven million miles of injury-free record in Phase 1B operation. Handicapped passengers are also able to access the PRT system, with elevators available in all stations. In addition, there is a spot on the vehicle for wheelchairs

Since the Morgantown PRT success in 1975, numerous visitors from all over the world have observed the operation. In light of cost, design concerns, and political situations, many decided not to continue the project. As for this PRT, it continues to perform well as an automatic guideway transportation system that is way ahead of its time. After almost half a century of operation, it has so far not a single fatality, which is quite outstanding. As well it has managed to get passengers to their destinations on time. The rides are free to all students. The annual operation cost is around $3 million. For passengers, the fare is fifty dollars per month, and for employees are sixty-three dollars per semester; these fares cover around 50–60 percent of the operating cost.

Looking at the size of operation, the PRT performance and the conveniences, it is really very reasonable. Meanwhile, it has been a showpiece of the city and helped its economy; in the 2000s Morgantown had the lowest unemployment in the country. However, after all these years of operation, the infrastructure has aged and requires more maintenance in order to prevent accidents. On November 30, 2016, two vehicles crashed between Beechurst and Walnut Stations, and on February 10, 2000, two passengers and students were taken to Ruby Memorial Hospital because a boulder dislodged from the hillside hit the vehicle. Though PRT has challenges, it is clear that it can offer sustainable solutions to traffic congestion as the urban population gets more dense.

After all these years of operation, if one compares automobiles, buses, and trains, the mileage that Morgantown got out of the PRT is quite significant. An average car or bus today cannot last so long. Especially with a car, a sizable amount of time is in a garage (9).

Challenges – Potential Opportunities

The United States, in the 1960s and '70s, focused on lunar exploration and the completion of the program. Afterward, they needed an alternative project. Consequently, then President Nixon in 1972 stated: "If we can land three men to the Moon 200,000 miles away, we should be able to move 200,000 people to work three miles away." In many ways, it was more difficult than it expected. According to a congressional study in 1975, the UMTA had not realized capabilities of the private sector working together with local transportation authorities. Consequently, they decided to terminate the deployment of the automated guideway technology. One positive outcome from the study was that the US government could take action to remove the barriers on automated guideways and facilitate PRT advancements by providing contracts and grants (Transit AG, 1975).

It was around the time when Mideastern countries boycotted the US, which resulted in substantial shortage of oil supply. The PRT concept was again being looked at by a large senior- design task force at MIT. Their published report, Project METRAN, covered most of the ideas of the PRT. At that same time, PRT system development was already under way at West Virginia, which resulted in enhancing the situation of development. Though it was way over budget, which discouraged many companies and foreign countries' organizations to get going, actually it is considered a great success because of its track record of performance throughout these many years. For the US, it was a great investment, since it is the first operating PRT ever built in the world, and to this day, many people are still learning as well as referencing it.

Though many nations and organizations have considered PRT, it entails a long payback time. The cost of trains and monorails are high: some $30–$80 million per mile. In recent years, new energy technologies have advanced and material prices have changed, which would sizably change the metrics of the price of the PRT systems as compared to those years. Consider the recent development of materials that are a lot lighter. The vehicles could weigh on the order of 400–600 pounds instead of over a ton, as in the case of an automobile. Modern infrastructure could also handle a number of functions which could provide better and shorter returns on investments. It would also modernize our aging infrastructure together with enhancing operation efficiency and longevity. It is time to move ahead and reassess the technology.

Apart from being serious polluters, oil and gas are depleting quite rapidly at the rate that they are being consumed. Knowing that all these resources are finite and despite legislation, there have not made a dent on solar installations. Recent years, there have been introductions on electric cars, but in looking at population growth, there have to be alternative solutions, and implementing new technologies to ameliorate congestion is at a serious crossroads.

Changing paradigms is always a concern; there are always hurdles to overcome, such as technical issues, economics, public sentiment, and legislation. These items take substantial time and resources to overcome. Any delays on one of these items could hinder the schedule of successful deployment and installation of the PRT system. Though a number of nations had done studies throughout the years, they have not taken serious action. Hopefully, this will change when more organizations come to participate and get the PRT system implemented successfully.

Climate change has already started to affect our lives, especially the coming generations, which all of us should be aware of. Droughts, horrific storms, forest fires, and extreme weather conditions that could be devastating to our food and water supplies, as well as more likely pandemics, e.g., Ebola, COVID-19, and SARS, are prospective problems. It is such an immense challenge that we should all work together to tackle it by implementing advanced transportation technologies like the PRT and abundant clean energy resources such as silicon, hydrogen, and oxygen as soon as possible.

A typical PRT vehicle can carry four to six passengers at max. However, it is much lighter than a car. Applying recent solar technologies could generate the energy to operate the PRT vehicles. If more energy is generated, it could either be stored or fed into the grid. The wonderful situation with photovoltaic systems, as long as they are well made, is that they could last twenty-plus years. In that case, energy is all being paid for in advance, which will reduce the schedule of payback or ROI (Return on Investment).

Energy, as we all know, is one of the major components in operating the transportation system. In this case, it also gives us the option of not relying on fossil fuel, which comes from hostile countries and involves concerns on price fluctuation in case of crisis or shortages, which is recently due to the war in Ukraine. These are all unnecessary anxieties, which could be avoided. Additionally, the price of solar has come down quite a bit in recent years. Now is the time for us to seize this opportunity. Autonomous vehicles or self-driving cars are gaining more popularity, and they have some similar features of the PRT because they are automatic and electric in operation. However, parking is a problem, which will continue to affect traffic congestion.

The comparison is shown below.

Comparison between Autonomous vehicle and personal rapid transit

Autonomous Vehicle	Personal Rapid Transit
Parking is needed	Parking is not needed
Upfront cost	Travel cost only
Traffic congestion unavoidable	Traffic congestion avoided
Cannot guarantee arrival time	Guaranteed arrival time
Personal maintenance required	Personal maintenance not required
Personal insurance required	Personal insurance not required
Chance of breakdown could be high	Low chances of breakdown
Personal fuel cost required	Personal fuel cost not required
Not require going to station	Require going to station
Chances of accidents	Unlikely chance of accident
Technically could have problems	Technically is much easier

There are concerns about job opportunities once the PRT is being implemented, due to the fact that it is an automatic system. Of course, some jobs will be eliminated, but other jobs will be created. We should try to adapt to it because this is how our economy grows: by making advancements.

Meanwhile, the new jobs created will be high-paying ones such as engineers, systems analysts, controllers, designers, managers, and architects. As the PRT system grows, it could spread to different locations that present their own requirements. Basically, PRT systems constitute a physical internet.

It was a wild dream that the cyber internet managed to create so many jobs and provided so much opportunities. Yes, the physical internet could create a lot of new jobs, but there will be some jobs that could reduce taxis, Uber, and Lyft. Meanwhile, traffic will be less congested and in return will be able to provide more rides once the PRT systems have been implemented.

The main advantage of the PRT is that it goes directly to the destination, but in the case of buses or trams, apart from making stops, one might have to make connections, especially in urban areas. Waiting time for the connecting vehicles could be long, and energy consumption is also a concern. Therefore, all these could be avoided in the case of PRT, especially during rush hours. Indeed, PRT could play a vital role to resolve traffic congestion because it reduces the number of vehicles going back and forth in urban and commercial areas as well as saving parking spaces. In tight city areas, planning stations could be quite challenging, though the required space could be limited. A well-designed layout of the stations is critical because it could provide better flow of traffic.

PRT in the World

System	Location	Guideway Length	Size	Feature
Morgantown PRT	Morgantown VW, USA	8.7 miles	5 station/73 cabin	20 passengers max
Cyber Cab	Masdar City, UAE	1.0 mile	2 station/10 cabin	Not operational yet
Ultra PRT	Heathrow Airport, England	2.4 mile	3 station/21 cabin	Terminal 5 & Parking lot
Skycube	Vectus	2.9 miles	3 station/40 cabin	

After all these years, the Morgantown PRT still takes the lead.

Latest Technologies – PRT Systems Applications

The PRT concept has been around for quite some time. Deploying the latest technology could not only solve the traffic congestion issue, it could also help resolve the problem of continuing reliance on fossil fuels, which are depleting quite rapidly. They are a major cause of climate change, and pollution related to fossil fuel extraction and use results in creating a very unhealthy environment, especially in densely populated urban areas throughout the world. Right now is an important time to reexamine the PRT system and apply the latest solar and storage technologies. Putting them together will strongly enhance the groundbreaking introduction of twenty-first-century energy and transportation technologies to the world. "PRT is a concept that has been evolving for over 50 years. With today's innovative technologies, PRT is emerging for the future because it has the potential for contributing significantly to solution to the fundamental problem of modern society including congestions, global warming and dependence on a decreasing supply of cheap oil," note Frost and Sullivan Analyst Poonam Tamania Dec. 2008 (14).

Applying latest solar technologies, especially BIPV (Building Integrated Photovoltaic), together with storage technology like fuel cells and batteries, could lead to a closed loop operation. An example is like collecting the solar energy and splitting water into hydrogen and oxygen. Hydrogen could then be stored and could be used in fuel cells on PRT vehicles. Green hydrogen is going to be more affordable because more nations are investing in manufacturing.

In addition, electrolyzers are getting more efficient and economical to manufacture. And PEM (Proton Exchange Membrane) Fuel Cells are more available. On the solar technology side, applying the BIPV products would also make a difference because the modules would also be part of the weather skin of the structure. They could form a cover or shade or façade. They could also be termed "active building materials and building products." Though initially there have been installation issues, advancements have been made to simplify the procedure, which will reduce time and costs as well as avoiding errors.

Looking at current transportation systems—for example, trains, subways, buses, and automobiles—they could be a potential hazard to riders during the prevalence of COVID-19 or other instances of widespread contagion. Especially in rush hour, distancing would be nearly impossible to implement. Recently a study by Prof. Jeffrey Harris at MIT found that the New York City subway is a major disseminator of the COVID-19 virus (7) and contagion in the London Tube (8).

Having masks on would be helpful, but social distancing could be a problem in a subway. PRT, which is basically self-driving, could avoid these issues. Climate change will make our planet problematic, and without drastic action to counter it, the place will soon be uninhabitable for a sizable portion of the population.

System Operation and Energy Storage on PRT

The growth of solar manufacturing and advancement simultaneous triggered advancements in storage technology. Though energy can be captured from photovoltaic panels, it is necessary to store the energy. Initially, lead acid batteries were used. Since the development of lithium-ion batteries, they have been applied to storing the energy from solar collectors. However, lithium-ion batteries can be hazardous. Meanwhile, currently the operating cycle of batteries is relatively low as compared to internal combustion engines and requires constant recharge.

Hydrogen could be another source of energy, obtained by splitting water, and fuel cells could generate the electrical energy for running the PRT. In addition, fuel cells have a much longer operating cycle, which is equivalent to or even larger than internal combustion engines. Therefore, hydrogen fuel cell systems are preferred. Subject to the design of the energy storage and operation system, either fuel cells or batteries could be installed on the vehicles. The storage unit could be recharged or refilled at stations when the vehicle makes a stop.

Ottobahn GmbH

Ottobahn –the emission-free, autonomous transportation solution that brings the soil back to people.

Ottobahn operation
Courtesy to Ottobahn GmbH

The Ottobahn transportation system carries people and goods in the new dimension, above today's traffic. It is made by hanging pods for up to 4 people running on proven railway technology.

Order your individual travel pod any time to any location within the Ottobahn network. The pods can be lowered to the ground everywhere along the tracks. The Ottobahn pods determine your individual route through our rail network and take you safely to your destination. The passengers will travel autonomously in a comfortable, electrically powered pod that picks them up on-demand at their door and takes them to their destination on time and emission-free.

Intermediate stops, changing means of transport or searching for a parking spot will belong to the past. Our pods drive with a speed of 60 km/h in the inner-city and up to 240 km/h for inter-city connections. Thus, a trip from Munich to Berlin could take only 2.5 hours. The efficiency of the transportation system comes from smart software. All pods coordinate their routes with each other with the help of AI. By using intelligent algorithms, the traffic flow in the Ottobahn network is optimized in real-time and controlled in an energy-efficient way.

The small and light Ottobahn pods are driven by efficient electric motors. The friction of the steel wheels is significantly lower than that caused by the rubber wheels of a car. This enables an energy consumption of the equivalent of only 0.5 L per 100 km. We also rely 100% on renewable energy for power supply. On the track, we will provide areas for solar panels. Therefore, we will be emission-free not only locally, but also globally.

The need for additional space on the ground for the poles is minimal. This allows Ottobahn to be easily integrated in existing infrastructure. In addition, the Ottobahn track itself and the area underneath it can be greened. Thus, the Ottobahn solution gives urban space back to citizens and nature, transforming streets into green lifelines for a better quality of life in cities.

Ottobahn GmbH was founded in July 2019 and is based in Munich. The team consists of highly qualified software developers and mechanical engineers. Since February 2020, the first test track has been in operation in the factory halls in Munich. The second generation of the power unit, the cabin, and the AI-based software control system have been completed. A real-size reference track will be built in the beginning of 2023 in the south of Munich.

We solve real traffic problems in all the world's major cities. In addition, the Ottobahn solution is cheaper and more efficient by a factor of 5–10 than any other means of transport (car, train, bus, etc.). We already have more than thirty leading cities interested in our solution. Ottobahn hits the nerve of the time and is a concept with a future!

Studies Done throughout the Years

Countless studies have been done throughout the years, but unfortunately, almost all of them, for one reason or another, result in not moving ahead on implementing the PRT system. Some of the reasons were political, involved economic issues, or misapprehended the assumed parameters of the study. Sometimes nobody wants to be the first to try it out in case the system fails to perform as anticipated, which could lead to a lot of finger-pointing. We should examine the system in Morgantown, West Virginia, as a model. As noted by the mayor of Morgantown:

- It has proven to be a reliable system of automated transit that is relatively inexpensive to operate.
- It has offered on-time service better than the bus system that it replaced.
- It has eliminated much of the gridlock of traffic that existed in the hub of downtown Morgantown.
- It has proven to be safe, with no serious injuries reported since the operation began in 1975.
- Approximately 16,000 riders take advantage of the system daily.

None of the companies that participated in building the system had any prior experience. It turned out to be an outstanding success story. Of course, to design a system applying the latest technologies could be a challenge. Potentially, there are risks on new development projects. As long as the project is well planned and designed, the chances of negative issues will be minimized instead of waiting somebody else to do it. If organizations back in the 1970s were able to do it, there is no reason why companies today cannot do it. To get started, the initial stage of deploying PRT could be on feeder rail on the existing transportation like trains and buses.

ConclusionAs we have noted, in the past couple of decades, there have been a number of experimental types of action on PRT around the world. In addition, there are a substantial number of case studies. Most of them believed the PRT would solve the traffic congestion issues, including cost saving on operation of the system. However, there have been hardly any real commitments on the PRT deployment since Morgantown except Ultra at Heathrow Airport, which only goes between the terminal 5 and the long-term parking lot. System price is quite expensive because the vehicle alone costs around 140,000 euros, which is not quite practical and affordable. Prices in the order of ten to twenty thousand dollars would be a more reasonable range.

In one of the case studies in New Delhi, India, the vehicle spacing was at eighty meters apart, which should not be a reality in an automatically controlled system, as was assumed; instead it should be below thirty meters or one second apart, and one meter apart should be the minimum. To get the PRT implementation started, let us look at a feeder rail system going between airport terminals and hotels, train stations, or parking lots, where one could book it on demand. If all went well, the project could be expanded to a larger system, or the PRT could be tried in more airports. It is wise to give a try, though there could be opposition. Once the traffic congestion is solved, all of us will be winners. Later on, larger and more sophisticated PRT could be implemented, which will be one of the ways to solve climate change.

References

1) "Dream today, reality tomorrow" Claire Marchand, Managing Editor, IEC, Geneva, Switzerland.
2) Anderson J 1996 Some lessons from history of personal rapid transit(PRT) In conference on PRT.
3) Anderson J.E. Some lessons history of personal rapid transit (PRT) 1996. Version 2 https://faculty.washington.edu/jbs/itrans/history. (accessed 2020)
4) Ramey, S Young Y.S 2005 Morgantown people mover update description
5) Anlauft A Frost & Sullivan. The Global Emergence of Personal Rapid Transit, London, December 2008 /PR Newswire/https://www.investigate.co.uk/frost-38-sullivan/Pm/the global-emergence-of-personal-rapid-transit-/20081201133100NN589/
6) https://www.youtube.com/watch?v=iaSaWfw07Sw U tube video on Morgan Town PRT 2015. Ref: 2022
7) The subway seeded massive coronavirus in New York City. Jeffrey Harris, MIT 2020
8) Analyzing the link between public transport use and airborne transmission, mobility and Contagion in London underground (L. Goece' and A. Anderson)
9) Wikipedia 2022 Personal Rapid Transit https//en.wikipdia.org/wiki/personal-rapid-transit (Assess 2022)
10) Transit AG 75 -Automated Guideway Transit Assessment on PRT by US Office of Technology at the US Congress.

Chapter 5

BIPV GREENHOUSE APPLICATION

Courtesy of Como Park Office

Food, as we all know, is vital to our existence, together with energy and water, which are required to grow food. According to the latest statistic on farmland, over 40 percent of arable land is being used in growing food, which is quite huge. As populations continue to grow, more land must be allocated to grow food.

Meanwhile, in light of climate change and land scarcity, a lot of places would not be suited for growing crops. Greenhouses are a major solution to alleviate the situation.

In 1845 Como Park was started in St. Paul, Minnesota, but not until 1914–15 did they construct the conservatory. Then, around 2000 that the McNeely Foundation decided to provide a sizable donation to expand the new wing of the conservatory. Later on, Atlantis Energy Systems Inc. was asked to build the translucent glass/glass solar panels for the greenhouse's new extension. The purpose of the modules was to provide partial shading on the Bonsai all year round. Three different types of modules allow quite a bit of daylight to come through, which is vital to the plants in a greenhouse. The system size is only 12 kw, which is relatively low. The energy probably only covers the lighting needs, which are subject to the design of the system. This is a great showcase applying building integrated photovoltaic technology in this landmark site. Como Park is supposed to be the second most visited place in the state.

Como Park Skylight

In the Como Park case, the greenhouse is mainly a museum for plants being displayed to visitors. However, when a greenhouse is used for food production, cost competitiveness is paramount. Nowadays, there are multiple levels of greenhouses that could grow different plants in one location. Sometimes a greenhouse may also grow breeding fishes. Tilapia and numerous other types of fishes could be bred in-house. Recently, the owner of a greenhouse of a couple hundred square feet managed to grow food plants on the upper level and tilapia on the lower level.

All conditions and requirements for setting up greenhouses are subject to the location. Applying BIPV and BIPVT in energy generations is the way to go. Why? Sunlight is critical to growing crops, along with getting energy to control the temperature indoors. Having the BIPVT to obtain thermal energy would add more energy to the greenhouse, especially in the cold climate of places like the Midwest and the Northeast and even some mid-Atlantic states. The thermal energy could be stored through tanks or geothermal systems. Having tanks could be simpler and less costly. A heat pump could handle the temperature control in the greenhouse: in the winter, the heat stored in the tank could be extracted, and in the summer, the thermal energy collected could be transferred to the tank.

If the greenhouse is located near a river, water could easily be accessible, which is a big plus on one of the key items. Sometimes, hydroelectric power could be developed so that apart from getting water, electrical energy could also be provided. But in case water is not easily available, one could collect rainwater and filter or purify it. However, if surface water is not available at all, a deep well may be required. If the location is close to the sea, desalination is another option, which could also employ the sun's energy to get it going.

Chapter 6

EDUCATION AND LOOKING AHEAD

Looking ahead, we all have a sizable task. According to the UN IPCC's most recent report on climate, we have less than ten years. Unless drastic action is taken in the near term, it could hard to reverse the current trends in climate change. As mentioned at the beginning of the book, Kenya, which is still considered an underdeveloped nation in Africa, is already close to being a net zero energy consumer. Especially the G7 countries, supposedly the most advanced nations on the planet, are not even coming close to it, which is really a shame. There is a saying, "Where there is a *will*, there is a way." Do we have the will? Yes, we have hurdles, but we have to jump over them to get things going. A major one is education especially in the solar technologies, because the world needs to reach 40 percent solar energy generation quickly, according to many experts. It is highly recommended to apply BIPV or BIPVT on houses and buildings. Listed below are some of the schools, such as the Manuel Elementary School, with installation of Megaslate roof tiles on campus.

Building Integrated Photovoltaic Installation from Elementary School to Universities

Manuel Elementary School	Bern, Switzerland
Kit Carson International Academy	Sacramento, California, USA
Kankakee Community College	Kankakee, Illinois, USA
Rock Valley College	Rockford, Illinois, USA
Utah State University	Logan, Utah, USA
University of British Columbia	Vancouver, Canada

Columbia University	New York, USA
MIT Climate CoLab	Massachusetts, USA

In light of the current urgent concerns on climate change, it is important to introduce the BIPV technology to schools and universities so that they may pass on firsthand knowledge to young learners. Hopefully, it will soon be a regular curriculum in schools throughout the world.

Manuel Elementary School – Bern, Switzerland

Megaslate covered the Bern, Manuel Elementary School (courtesy of 3S Swiss Solar Solutions AG))

The initial construction of these school buildings was fifty years ago, and installation of the solar roofing system allows them have their electricity needs met for at least twenty-five years. The Megaslate, a larger format module, manufactured by 3S Swiss Solar Solutions, also has a reliable track record and would be a good choice for expansive roof planes.

The 590 kw system is the largest BIPV installation on a school in Bern and covers 3,500 square meters. Currently in Switzerland, hydropower supplies approximately 60 percent of electrical energy, while 5 percent is from solar power. They hope to meet the net zero energy goal by 2050. Right now, the school supplies 40 percent of the total electrical energy. The Manuel School was awarded the Swiss Solar Prize on renovation.

The Manuel Elementary School in Switzerland and Kit Carson International Academy in Sacramento give students a head start in becoming aware of building integrated solar technology, which can power the buildings, and it would be beneficial for them to appreciate all the advantages it offers.

Kit Carson International Academy – Sacramento, California, USA

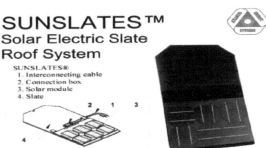

Kit Carson International Academy – 33 kw system at Sacramento Municipal Utility Municipal District (SMUD)

In the late nineties SMUD pioneered the solar program in Sacramento, California, which offered very attractive incentives on solar installation. The program drew a lot of attention from homeowners as well as major home builders such as US Homes, Beazer Homes, and Morrison Homes in the district. Kit Carson Middle School participated in the program around 1998. At that time, Atlantis Energy Systems, Inc., had just developed the first BIPV solar roof tile. SMUD committed to an agreement with the company to install the Sunslate in the district. They led the way among utilities on BIIPV applications on rooftops, where Atlantis was able to install over two hundred homes in their location. Unfortunately, the program cost the organization a substantial amount, which resulted in making changes on the program. On Sunslate, though there were some hiccups at first, as with most cutting-edge products, it turned out that to this day, after twenty more years, it is still operating well on all the installations.

Kankakee Community College – Kankakee, Illinois, USA

OWNER: Kankakee Community College , Fine Arts Building
SYSTEM: Curtain Wall
POWER: 2 kW
LOCATION: Kankakee, Illinois

Kankakee Fine Arts Building: Insulated Glass/Glass Curtain Wall
(Courtesy of Kankakee Community College)

Two community colleges in Illinois installed insulated curtain wall. Kankakee started off by installing the insulated glass, and six years later, Rock Valley decided to take action on installing the insulated glass but included other supporting systems, which will save a lot of energy.

The installation at the Kankakee Community College took place in 2006. The solar part was only around 2 kw. In those days, geothermal storage and heat pump technologies were not quite ready yet. To obtain more sunlight on buildings, it is both attractive and efficient to apply insulated glass along with photovoltaic. Although in this case, the photovoltaic portion is quite minimal, it could be helpful to tie in on different systems in the building. That is why it could only support the operation of the toilets in the building at Kankakee. However, in the insulated glass situation, in summer and winter, it shields off heat and cold, which saves energy consumption. In Kankakee, the insulated glass uses air to fill as insulator while at Rock Valley, the glass uses argon gas which is the best gas insulator available.

Rock Valley College – Rockford, Illinois, USA

Rock Valley insulated glass current wall (Courtesy of Rock Valley Community College)

Apart from the BIPV curtain wall, the Rock Valley system consists of the items below:

1. 420 ton closed-loop geothermal heating and cooling
2. Water-cooled heat pumps
3. Chilled beam cooling system
4. Energy recovery on air handling system
5. Variable volume heating and cooling pumps
6. Thermally efficient glazing (low-E coatings, ceramic fritting)
7. White membrane roof W/R Insulation
8. Geothermal domestic water heating
9. Variable frequency drives
10. High-efficiency lighting fixtures
    - Daylight/occupancy sensors
    - Multiple switching zones
    - Daylight harvesting sensors
    - Use of low mercury lamps

11. Variable frequency drive for motors
12. Enhanced commissioning

All the above information is important knowledge for students to learn so that they can apply it after completing their education.

The Rock Valley Community College completed the Karl J. Jacobs Center of Science and Math in 2012. It was awarded LEED Gold by the US Green Building Council. As one can see, the whole system in the building has a substantial number of advanced energy-saving systems. A geothermal system also plays a major role. In this case, solar electric not only produces the energy, it allows sunlight to pass through, which brightens the interior as well as making it aesthetically pleasing. The system has an annual saving of $76,000. In light of recent developments on storage and generation technology systems, the energy and cost savings could be higher.

Normally, insulation with glass/glass solar modules ends up to be quite heavy. However, this is like triple glazing: it very much lowers U values, which is good for keeping warm air in buildings as well as blocking noise in busy locations. In places like the state of Illinois, where winters are quite severe, the triple glass windows could save substantial amounts of energy and air conditioning cost in the summer months.

Utah State University – Logan, Utah, USA

The solar power system at Utah State University consists of two separate arrangements. On the rooftop are the regular standard modules, and at the south-facing part of the building are the BIPV modules. These are shades which diminish energy consumption in the summer as well as generating electrical energy. The system consists of 108 modules, each of them at 310 watts for a total of around 33.5 kw. At that time, it was supposed to be the first and the largest BIPV system installed around the Cache Valley area. The deployment of the cutting-edge and energy-saving technology impressed students and offered an opportunity for them to learn that solar installations could be aesthetically pleasing as well.

On top of the advanced solar system, the building is designed to optimize efficiency of energy and water consumption. On the construction side, it also uses the latest advanced building material to compensate the system, which resulted in winning the Gold LEED certificate for the Stan L. Albrecht Agricultural Science Building.

University of British Columbia – Vancouver, Canada

Centre for Interactive Research on Sustainability (CIRS)

This is a laboratory to showcase building design for sustainability and conduct research as well. The project was conceived by Prof. John Robinson of the University of British Columbia with the architect Peter Busby of Perkins and Will. Their goal was to be the most innovative high-performance building in North America. Project design started in 2008, and the site on the UBC campus in Vancouver was selected.

The total size of the building is around sixty thousand square feet. The BIPV system with the skylights and solar shade modules covers around 25 kilowatts, around 10 percent of the total electrical energy. As to the solar shade, apart from generating electricity, it is being used to control glare and heat gain. The rest of the electrical energy comes from British Columbia's hydro grid. For domestic hot water, they installed an array of evacuated tubes on the rooftop to capture the solar thermal energy.

The building has its own system to collect rainwater at all the rooftops, which makes it totally water self-sufficient. The water goes through a filtration system and is sanitized on site prior to being distributed to different parts of the building. In addition, the indoor air ventilation system is being monitored to control and maintain air quality. As to lighting areas, the building is designed to optimize the use of daylight to reduce the use of artificial indoor lighting as much as possible (1).

The project applied IDP (Integrated Design Process) to assist on the design, and the team worked closely to make it happen. In addition, the team used the BIM (Building information Modeling) software to layout the drawing plans and construction documents, which resulted in developing a very well thought-out design.

The prime objectives of the facility are to serve as an interdisciplinary hub as well as a sustainable building research facility which stands as a living laboratory on campus. For anyone visiting Vancouver, visiting the center is highly recommended. One could learn a lot about how sustainability can be applied.

The project was awarded the first LEED Platinum certification of the Living Laboratory.

2012
- Architectural Innovation Award: Architectural Institute of British Columbia (AIBC)
- Award for Engineering Excellence: Association of Consulting Engineering Companies
- Excellence in Structural Engineering Award: National Council of Structural Engineers
- Perkins + Will Design Biennial – Design Merit Award

2013
- Sustainable Building of the Year: World Architecture New (WAN)
- Sustainable Development Award: Golder Associates

2014
- Canadian Green Building Award: SAB Magazine

2015
- ISCN 2015 Sustainable Campus Excellence Award, Excellence in Building:
- International Sustainable Campus Network
- Green Building Award: Royal Architectural Institute of Canada (RAIC)

In addition, there have been another seven awards.

Columbia University – Earth Institute – New York, USA

The Earth Institute was founded in 1995 and currently has sixty to seventy full-time members. In addition, hundreds of interdisciplinary faculty participate in a substantial number of programs, from environmental science to sustainable materials and energy harvesting technologies, to architecture and political issues, as well as legal concerns between countries. Recently, they have also founded the Climate School. A great part of the Earth Institute is located on the Lamont campus which is across the Hudson River. As a whole, it is probably one of the largest institutions in the US to tackle climate change.

MIT Climate CoLab

The Climate CoLab is part of the Center for Collective Intelligence at MIT, and the objective is to gather all the intelligence of tens of thousands of people, especially scientists, all over the world to handle the complex societal problems linked to global climate change. It is an open problem platform of a growing community of over 120,000 persons, including leading international experts on climate change and relevant parties such policymakers, business investors, and concerned citizens, including the United Nations. They participate to gain momentum in achieving global climate change goals. Recently, the Climate CoLab has joined with InnoCentive in widening the operation.

In light of education, more schools and universities have started to offer programs in climate change studies. It is vital to have the workforce ready to race against time to meet the challenges. However, in addition to education, working with government organizations and major corporations are other major hurdles that should be looked into. Though advanced technologies present many issues on development and commercialization, as long as the potential end results in the appropriate solution, it should be continued. An example is green hydrogen development, which has been very challenging throughout the past two decades; it is now starting to become a reality. The important end results are (1) no pollution, (2) abundance, and (3) resources not dependent on a limited number of countries. It will probably require a couple of decades to build up the infrastructure. The time to do it is now.

On a positive note, Denmark has committed with Toyota to get one hundred of their Mirai hydrogen cars for their taxis in Copenhagen, up to two hundred by year's end, and up to five hundred by 2025. The Toyota Mirai broke the record with a run of more than a thousand kilometers without a refill in France (5), and the latest news is that most likely Paris will be getting fifty Toyota Mirais for taxis.

Japan and China are going strong on hydrogen cars, but unfortunately, China is still highly dependent on coal instead of renewables. On the EU side, a sizable number of countries are investing in hydrogen fuel as energy; Germany and France are taking the lead, aiming at 7–8 GW and 3–5 GW respectively by 2030. Most European countries are also looking to domestic and industrial applications in addition to transportation. On transportation, Audi and BMW will be coming out with hydrogen cars as well (12). As to air transportation, Airbus is planning to develop hydrogen engines for their jet planes. Hopefully, they can come up with it by around 2035. In the US, the Biden administration has provided $504 million through DOE that would be awarded to Mitsubishi Power Americas Inc. together with Magnum Development LLC, along with the private equity investment firm Haddington Ventures LLC. The project will also involve work from entities Black & Veatch, NAES Corp., WSP Global Inc., and the Utah School and Institutional Trust Lands Administration (7). In addition, the administration also awarded Monolith Inc. a loan guarantee of $1 billion to develop green hydrogen (7,8).

On the negative side, large fossil fuel and utility companies are taking strong action and finding ways to prevent renewable energy from being deployed and installed. Though many utilities have installed solar and wind facilities, they have taken up a substantial amount of precious land. The best way is through individual homes and buildings (because the space is already there) or local microgrids, which would allow more local energy independence.

Besides, energy transmission is quite costly. A sizable amount of energy is lost through transmission, and the lines are expensive as well. Ten or fifteen years ago, line cost was already at around $3 million a mile. We should expect a reasonable increase above it today. Other options could be on top of highways and sound barriers on highways and railway tracks. Meanwhile, our grid is vulnerable, especially through severe storms, and our adversaries hacking them. Look what happened in Texas in February 2021. Some ten million locations had power outages and cost almost $200 billion, which was horrific, especially in the winter. We can expect more failures to come, and we should be prepared. To all sectors, however, in light of our current planetary situation, it is paramount to take a closer look.

What good is profit, if we don't even have a place to inhabit, especially for our grandchildren? Some people from the solar industry have mentioned this to me for the same reason. On top of that, many individuals of lesser means have worked tirelessly in the past decades on their concerns that their neighborhoods would be flooded due to rising sea levels. A lot of coastal cities will be flooded. As the Pentagon said a couple of decades ago, climate change is the greatest threat to our national security. It is getting clearer by the day. How come, after all this time, we are still holding back and not taking more drastic action? Right now, we are a bit late, but we can still catch up on it before it is too late. Wealthy corporations and individuals should look into making sizable investments in renewables like solar and wind.

On the solar side, BIPV and BIPVT is highly recommended because the technology could make energy generation more efficient and sustainable. Energy like green hydrogen is also highly recommended, especially because every nation can generate their own energy and become energy independent. Silicon processing operation is another one that all countries should look into. Hydrogen and silicon are going to be major energy resources in the future. To all government organizations, please take more drastic actions to pass legislation to provide more incentives to reward individuals and companies that install or develop renewable technologies.

Certain fossil fuel companies recently made presentations. One of them mentioned that they are planning to reduce their production of oil by 40 percent by 2030 while developing renewable technologies at the same time. Meanwhile, they have turned over employment by hiring new workers on renewables as well as retraining some current employees. In addition, hydrogen was mentioned quite a bit instead of batteries, which is great.

Despite these sounds of hope, we should all take appropriate actions by installing more solar panels on our homes and buildings on our own initiative because time is of the essence. Also the federal incentive is great this year.

A recent *New York Times* article indicated the 1.5 degree C. rise in temperature could affect more than a billion people worldwide because the heat could be fatal to them. Meanwhile, a substantial number of places will have serious drought problems, which lead to food farming issues and wildfires. The rate at which climate is changing is harming the earth faster than we are adapting. There is flooding in China, Germany, and low-lying Asian countries. All these situations are creating famine and malnutrition as well as diseases like malaria and dengue fever, which are spreading to more areas through mosquitoes. Countries that have been taking more preventive measures have managed to reduce the damage (14).

It is time to take drastic action in reexamining our ways of constructing buildings, of farming, and of energy usage. Governments and major corporation should take all appropriate actions, including helping poor nations because they are more vulnerable. There was a pledge of $100 billion by wealthy nations, but so far, they have fallen short on their commitment. It is time to take action. However, on a positive note, humanity has managed to save a lot of lives from horrific storms and diseases, but investing in advancing weather prediction technologies and public health practices is crucial to provide advance warnings. But more has to be done if we want our planet to be saved from all these catastrophic incidents affecting all of us (14,15).

When COVID-19 was classified as a pandemic by WHO in spring 2020, there was quite a bit of talk about whether it had any ties with climate change. At that time, it was somewhat of an educated guess. Recently, an actual link has been shown to exist between climate change and the virus. In an article by Ed Yong of *The Atlantic*, he stated, "We created the 'Pandemicene' by completely rewiring the animal virus. Climate change is creating a new age of infectious dangers." According to estimates there are on the order of forty thousand viruses being carried by mammals. A quarter of them could affect us.

As the climate changes, it forces animals to relocate themselves, resulting in newly neighboring species. Some of them could be carrying infectious viruses. As these species find more attractive habitats and they also mingle with each other as well as getting closer to humans, these viruses could eventually infect us. In light of the current situation, it would be impossible to reverse the trend; perhaps it could have been possible four or five decades ago. Right now, public health advancements could make us more alert to take drastic action to protect us. More investment in these areas is highly recommended so that we are better prepared for the next virus.

We all know that the sun has a tremendous amount of energy, but do we know how much? According to Prof. Kaveh Vasei, an astrophysicist at the University of California, Riverside, stated in November 2020 that the sun's generation of energy in less than one hour is equivalent to the energy that our planet earth consumes in one year. This is very powerful information (11). Back in the early 1900s, when Einstein theorized the implications of the speed of light, it was an important one to space exploration and especially satellite communication systems, which have recently brought a lot of conveniences into our daily lives. Now that we have learned that the sun is so powerful, we should look into directions to capture as much clean energy as we can on our planet earth and take care of all our needs.

Since the water supply is already a serious concern, considering the sun's energy, especially in coastal towns, cities, and states, water desalination systems applying solar power should be looked into. In addition, thinking of the recent heat waves happening in North America, Europe, and sizable parts of the world suffering in extreme weather, applying the modular BITERS system could be one of the major solutions. Right now, it is still in development. Hopefully, around spring 2023, it will be ready for manufacture.

Solar power technology is at its infancy, as computer science and semiconductor industries were back in the early 1970s. It is now up to us to seize this opportunity to all work together and make all these new clean technology products into a reality that will make a better world instead of fighting among ourselves to destruction.

As mentioned earlier, although silicon is abundant, quality is paramount because the initial investments on converting from sand to silicon wafers is awfully high, in the tens of billions. The quality of the silicon module is mainly based on the final process which is the least expensive part of the whole process, amounting to around 10 percent of the cost. Module manufacturers could invest in better material and manufacturing machinery, which could improve the quality.

The silicon and wafer cost in terms of percentage would around $0.60–0.63, and the solar cell and module $0.37–0.40. The module manufacturing alone would cost $0.10–0.12. If the module could last 20–30 years or longer, it is well worth it.

Construction of houses and building is also a very important point. The effects of hurricane Maria in Puerto Rico five years ago are still problematic, and now after hurricane Ian in Florida, the devastation in Fort Myers Beach and Pine Island is horrific. In light of the situation, please look at the *Economist* article on how Hong Kong managed to handle Mangkhut, the strongest typhoon since World War II at wind gust 260 kph back in September 2018. There was not a single fatality and only 363 injures. One of the main reasons was that the buildings were built to withstand the hurricanes. This should be the norm starting right now, especially on rebuilding (16). We could do the same; why don't we? Looking ahead, our main objective is solving climate change. It is a tremendously large problem to tackle. Not a few organizations or countries could handle it. Last century, we fought two world wars, and now we are fighting a common enemy which incidentally is of our own creation. That is why we have to all work together in all fronts if we are going to win this war for our children and the coming generations in order to preserve our planet, Earth.

As time is running out, how can we approach it? Right now, there are only eight years left to make major changes. Afterwards, according to UN IPCC, we will not be able to reverse the degradation of the climate. We have to take action at once and set a mandate.

There is a personal experience learned while being at IBM. In the early 1960s, the company announced the model 360 computer system, the first multiple input and output system, which would substantially increase the performance of the computer. However, there were serious technical problems, especially the software part, while the company had invested heavily in the project.

The company was at a serious crossroads. If IBM could not deliver the system, it would be the end of the company. Senior managements were not in agreement with each other. Consequently, they were invited to the IBM Homestead in upstate New York and told to sort out their differences within a certain period of time. Luckily, they were able to sort out the situation and decided to work together.

The 360 system was a big hit, and the company managed to capture the largest market share, especially in mainframe systems. In the 1970s, IBM continued to grow after they started delivering the 370 system and became the cash-richest company, surpassing General Motors at that time. The moon landings and the Space Shuttle flights were also handled through IBM computers. After six decades the IBM mainframe is still being used by Fortune 500 companies as well as other large institutions. A lot of the world's scientific achievements would not have been possible without it.

Working together with a mandate and being transparent with each other would give the opportunity to get to know ourselves better, which would result in moving ahead successfully. Just imagine how, once we have slowly managed to stabilize the climate conditions by applying renewable energy resources, solve the water and food problems, and control virus transmission, we can all live a healthier life peacefully despite being more populated. It might sound a bit idealistic. Let's give it a try, and perhaps the world will be a lot richer than ever. Our health is also our wealth. Right now, the world is in a similar crossroads situation regarding climate change. Therefore, we should all work together for the sake of the coming generations and give our beautiful planet earth a chance.

"Time and tide wait for no man" (Geoffrey Chaucer, fourteenth century).

"Climate change waits for no man and respects no borders. Let us all unite as one and tackle it. The time to act is now, and we cannot afford to postpone."

1) Reference: UBC CIRS building manual 1) Energy System section, 2) Water Re-claim section and 3) Day Lighting System section.
2) https: novergysolar.com/understanding-the-degradation-phenomenon-in-solar-panels
3. The subway seeded massive coronavirus in New York City. Jeffrey Harris, MIT 2020

4) Analyzing the link between public transport use and airborne transmission, mobility and Contagion in London underground (L. Goece' and A. Anderson)

5) https://renewablesnow.com/news/toyota-everfuel-drivr-tie-up-for-more-hydrogen- fuelled-taxis-in-denmark-768341/ Summer 2022

6) https://www.intereconomics.eu/contents/year/2021/number/6/article/green-hydrogen-in-europe-do-strategies-meet-expectations.html Summer 2022

7) https://www.eenews.net/articles/doe-unveils-500m-loan-for-massive-clean-hydrogen-project/

8) https://www.latimes.com/business/story/2022-01-24/biden-revives-clean-energy-program-with-1b-loan-guarantee

9) https://toyotatimes.jp/en/toyota_news/163.html

10) https://www.nytimes.com/2022/02/28/climate/climate-change-ipcc-report.html

11) https://www.wyzant.com/resources/answers/795303/how-long-does-it-take-the-sun-to-deliver-to-earth-the-total-amount-of-energy.

12) https://www.press.bmwgroup.com/global/article/detail/T0403302EN/bmw-group-commences-in-house-production-of-fuel-cells-for-bmw-ix5-hydrogen-in-munich

13) https://www.pv-magazine.com/2022/05/09/latest-eu-electrolyzer-pledge-could-speed-up-solar-permitting/

14) https://www.nytimes.com/2021/08/09/climate/climate-change-report-ipcc-un.html

15) https://www.nytimes.com/2022/07/23/opinion/biden-climate-change.html

16) https://www.economist.com/international/2018/09/20/in-hong-kong-common-sense-keeps-lives-safe-in-a-typhoon

Glossary

Abum	Active building material ™ registered trade mark e.g. solar roof
Abup.	Active building products ™ registered trade mark e.g. solar roof tiles
AIA	American Institute of Architect
BIPV	Building Integrated Photovoltaic is a weather skin of the building that generate electrical energy
BIPVT	Building Integrated Photovoltaic Thermal is a weather skin of a building that generate electricity and thermal energy
BITERS:	Building Integrated Thermal Roofing System
FEW	Food, Energy and Water
Heat Pump	A system that can handle as air conditioning in the summer and heating in the winter
HERS	Home Energy Rating System- NET ZERO ENERGY – at 0 and the higher it goes, the home is less efficient and as the score goes up to 150. It is set by RESNET (Residential Energy Service Network).
HUD	US Department of Housing and Urban Development
HVAC	Heat, Ventilation and Air Conditioning systems
ICF	Insulated Concrete Forms
LEED	Leadership in Energy and Environmental Design set by the US Green Building Council (USGBC)
PCM	Phase Change Material - substances that change form while releasing energy being stored
Photovoltaic	A power generating devices that absorb energy from sunlight and convert them into electric energy

PRT	Personal Rapid Transit-an overhanging automated transit like a cable car or an automated taxis in the air
SMUD.	Sacramento Municipal Utility District

Synchronous communication – connecting at real time like phone call or video

Asynchronous communication – connecting at one's convenience

UMTA	US Department of Urban Mass Transit Administration
Umwelt Arena	It is an interactive museum on renewable technology in Switzerland
UBC CIRS	University of British Columbia - Center of Interactive Research on Sustainability
UN IPCC	United Nation Intergovernmental Panel on Climate Change
VOC	Volatile Organic Compound

About the Author

In 2001, Chairman/CEO Atlantis Energy Systems, Inc., and deploy BIPV(Building Integrated Photovoltaic) technology worldwide. Pioneer in the development deployment of the BIPVT(Building Integrated Building Integrated Photovoltaic Thermal) technology worldwide in 2007 that led the company awarded the World BIPV Niche Player of the Year in 2009 by Frost and Sullivan. Since then, advancements of the BIPVT have been made. In 2017 to present, Owner/CEO of Sustainable Product Development Lab (SPD Lab). Past number of years, was the Chairman of Columbia University, IAB(Industrial Advisory Board) in CEHMS(Center of Energy Harvesting and Material Systems) organization and board member of numerous companies. In 2019, SPD Lab and (AGP)Aesthetic Green Power Corp organization awarded as one of the top ten companies in the US in solar technologies by CIO Insight. SPD Lab holds numerous US patents and abroad. Co-authored with Prof. Yin and Mehdi Zadshir, Building Integrated Photovoltaic Thermal: Fundamentals, Design and Applications published by Elsevier Academic Press, 2021.

Education: MS EE 73   Tufts University
Experience: Over 30 years in semiconductor started with IBM, E. Fishkill NY in 1974

Printed in the USA
CPSIA information can be obtained
at www.ICGtesting.com
LVHW070845081124
795688LV00083B/673